DATE DUE

Demco, Inc. 38-293

0 9 2009

HELL
ON
EARTH

DAVID L. PORTER
with Lee Reeder

A TOM DOHERTY ASSOCIATES BOOK
NEW YORK

HELL ON EARTH

THE WILDFIRE PANDEMIC

HELL ON EARTH: THE WILDFIRE PANDEMIC

A Forge Book
Published by Tom Doherty Associates, LLC
175 Fifth Avenue
New York, NY 10010

www.tor-forge.com

Forge® is a registered trademark of
Tom Doherty Associates, LLC.

ISBN-13: 978-0-7653-1380-5
ISBN-10: 0-7653-1380-4

First Edition: August 2008

Printed in the United States of America

0 9 8 7 6 5 4 3 2 1

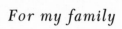

For my family

CONTENTS

AUTHOR'S NOTE

This book is about wildfire, and how fires affect us not only locally, but on a global scale as well. The effects of global warming reveal a direct connection to the unusual and unprecedented incidents of large-scale wildfire activity, often in areas that have never been prone to wildfire concerns until now.

But the bigger subject this book explores is the apathy that seems to go hand in hand with the disasters the wildfires, and their parents, global warming, continue to spawn. We can create fire suppression methods, and we can clean up our environmental act, but how can we win over apathy?

The ostrich, besides being one of the most unusual-looking animals on Earth, employs one of the quirkiest defense mechanisms imaginable: When the big bird fears danger in the air, he searches out a hole and sticks his head into it. Once safely ensconced in his cozy dirt dugout, the noble ostrich believes all is right with the world.

We certainly rival our friend, the ostrich, in our capacity to avoid any meaningful discourse on global warming. But by avoiding the truth and ignoring the facts, we have made ourselves as vulnerable to the effects of global warming as an ostrich is to a lion.

We live on a planet that, in every way, is a living, breathing, fully functioning organism that doesn't require humanity to get it through the day; the Earth has done quite well on its own for eons. Yet even though part of the world just as the plants, animals, and oceans are, we have proven ourselves very slovenly houseguests indeed.

In the last century human progress has brought about smoky factories and the ubiquitous internal combustion engine. Those two particular factors alone have done so much to encourage and expedite global warming, in such a brief span of time, that scientists question if the clock will run out before we fully understand the scope of the mess we created.

What is also interesting is mankind's ability to polarize along political lines even though such a monumental event, should it occur as scientists believe it will, would spell the end of life as we know it.

Global warming is the big picture, but people are usually coaxed into action only by events that affect them directly. After all, to most people, today looks exactly like yesterday, and it's difficult to imagine an ozone hole tomorrow.

Throughout this book you will read numerous accounts of fire safety experts who, for years, have sounded the warning bell. These fire professionals knew what to expect and, just like the scientists who continue to caution us about the effects of global warming, did and continue to do everything in their power to protect us. They wish to make us safer through preparedness by defeating the mind-set of apathy in regard to fire readiness.

In the fall of 2003 we were tested, and we failed. Many lost their lives, and millions of dollars of property vanished in an instant, as devastating firestorms pushed through Southern California.

The scene has been repeated many times since—for example, in the states of Georgia and Florida. Thankfully those fires didn't reach many homes, and loss of life was minimal, but the damage to the environment was done, and it will take decades before much of it will be restored, if ever.

As you read this book, consider that the fires that raged in 2003 (and are profiled in great detail), and after, were a harbinger of what would come nearly four years later to the date. In October of 2007 the greatest series of wildfires ever recorded struck the same general areas as the fires of 2003. And as unbelievable as it may be, the 2007 fires did even more damage. The 2007 fires were the result of keeping our heads buried snugly in the sands of ignorance. We can no longer afford to ignore.

The timeline in *Hell on Earth: The Wildfire Pandemic* is straightforward with the exception of the days between October 25, 2003, and October 31, 2003, when the narrative jumps among four simultaneous fires.

Also, all of the names of the people portrayed in the book are real, with the exception of certain fire officials, government employees engaged in undercover arson-related investigations, or witnesses to acts of arson where the case(s) remain unsolved or otherwise remain open. Changed names are indicated with asterisks in the text and the index.

INTRODUCTION

The firestorms that raged through Southern California in fall of 2003 destroying homes and lives should have been a wake-up call for the world, but they were not. Nor were the deadly and destructive winter and spring wildfires in Texas and Oklahoma in 2005–2006, or the Esperanza Fire near Palm Springs, California, which claimed the lives of five U.S. Forest Service firefighters. Nor were the fifteen hundred or so wildfires that, in 2005, destroyed vast areas of Brazil's Amazon rain forest—often called the lungs of our planet.

Except for the people and agencies involved in those horrific events, these fires are largely forgotten.

The firestorms of Southern California in 2003 are covered extensively in this book because I can recount them personally. I lived through them in another life that is behind me now in ashes. However, they only illustrate the horror and the challenge facing us. As we were to learn in the fall of 2007, the fires of 2003 were all but forgotten and certainly overshadowed by the fires that raged through Malibu, San Diego County, Lake Arrowhead in the San Bernardino Mountains, and areas of Ventura and Orange counties. And as huge as those fires were, they will soon be remembered as but a small taste of what is in store for the world as our climate continues to warm.

The people who lived through the 2003 and 2007 conflagrations in Southern California have not forgotten how it was. From San Diego County to the Southern Central Valley, the choking smoke was inescapable, even indoors. It labored one's breathing and permeated clothing and everything else it touched. It was as if one no longer had a sense of smell, because the only thing to smell was smoke.

Sirens pierced the air night and day, and traffic lights at many intersections were dead. It seemed there were more emergency vehicles on the roads than passenger vehicles. Electricity was the exception rather than the rule in homes and businesses. Many people were temporarily homeless, and during the 2007 crises, over a million people were temporarily displaced. Fear was palpable and it was everywhere.

By late October 2003, Southern California appeared to be under attack. It was really the first time so many fires had raged simultaneously in the urban wildland interface.

On a clear day, from the ridges of the San Bernardino Mountains, known widely as "the Rim of the World," one can see nearly this entire region, from the ocean to the desert and down into Mexico. During the fires of 2003 this area was a scene of devastation. From the ridge to the horizon, one mushroom cloud after another rose thousands of feet into the air, starting black at the bottom and exploding into massive white plumes at the top.

It was truly Hell on Earth, and hell returned in October of 2007.

This book seeks to highlight what happened during those weeks to point out the danger that everyone everywhere will one day face. Why? Because Hell on Earth is starting to happen now in many places that were once thought to be the last place such fire events would ever occur. A recent example is Greece where, in August 2007, ten wildfires raged at once in different parts of the country.

A greater disaster is going to come sooner rather than later, and it is difficult to imagine where you live if you believe it cannot happen to you.

Unfortunately, as we saw with Hurricane Katrina, our own government, like others around the world, seems to take preparedness seriously only after a disaster takes on almost biblical proportions. Will the firestorm that finally awakens the politicians happen where you live?

FIRES ON THE INCREASE

More and more the world is burning, and more and more it appears that we are to blame.

Devastating fires are increasing throughout the western

and southern United States, the number of fires in the Brazilian rain forest last year tripled, and vast areas of the wilderness are dying throughout the West, setting the stage for a human and environmental tragedy.

Conditions that create large-scale fire disasters are occurring more frequently every year, spurred on by global warming. And the potential for damage, loss of life, and greater harm to the environment is staggering.

The scope of the fire disaster in Southern California in the fall of 2003 and again in 2007 should have started a firestorm of legislative action in this country, followed up by real progress on the ground. It didn't. News reports of 2005 Texas fires told of the charred bodies of people who had jumped from their cars in a last desperate run for their lives, only to be engulfed by flames, and of cattle found by the hundreds, dead from the flames.

After enduring a deadly fire "season" that officially lasted nearly two years, real action should have been spurred by events in Texas. It wasn't. The story died in the headlines less than a week after the fire season was declared at an end.

Again, one need only look at the federal government's response to Hurricane Katrina to imagine what will happen when a massive firestorm hits somewhere in the United States. We are not ready for it, and it may happen where it is least expected.

People who live in major cities and who feel safe from fire need only consider what happened in San Bernardino, California, on October 25, 2003, when the Old Waterman Canyon Fire roared across the wildland-urban interface, fanned by sixty-mile-an-hour winds, destroying hundreds of homes in a matter of minutes. Tens of thousands of people who thought they had nothing to worry about just a half hour earlier found themselves suddenly fleeing with nothing but their lives.

Is this an isolated incident?

Consider what happened in Oklahoma and Texas late in 2005, when "perfect storm" conditions for fire prevailed over a large region for weeks and even larger urban centers such as Oklahoma City were threatened.

And the problem there only worsened. From late October to early March, most of Oklahoma had received only about one-quarter of an inch of rain. During that period, Tulsa should have received about 13 inches of precipitation. On March 1, 2006, in what should have been the beginning of the rainy season, arson caused grass and brush fires that destroyed more than thirty homes, raced largely unchecked across thousands of acres, and injured seven firefighters. On that day the high temperature was a record 93 degrees in Tulsa, with strong winds and humidity in the teens. The average high for March in Tulsa is 62 degrees, and the historical average number of days in March with a high above 90 degrees is *zero*.

Of additional concern: The devastating drought in the southeast is turning the normally green Georgia pines into matchsticks. What will happen when firestorms hit an area that has little water to fight a fire?

Also on March 1, 2006, a fast-moving fire in Colorado consumed thirty-six square miles of prairie. If these incidents can start happening in March, what will begin to happen in August and September?

We are going to have to live with this danger. Experts say that we have pumped so much greenhouse gas into the atmosphere that even if we stopped altogether today, the Earth would continue warming for many decades.

Apparently, fire begets more fire.

Global warming could be worsened by fires that devastate the tropics because the rain forests around the world serve to soak up carbon dioxide from the atmosphere. Brazilian officials have also speculated that smoke from drought-caused fires last year may even have made the drought worse by impeding the formation of rain clouds.

With this potential tinderbox forming, we face another threat—the opportunists who will take advantage of these conditions to create havoc and satisfy their depraved desires. In this book I profile the arsonist and give a detailed look inside that mind.

In addition to the human and economic toll of fires, there can be a great environmental cost as well. Throughout our country and in other developed nations, many endangered species cling to ever-narrowing areas of habitat as urban sprawl encroaches. In some areas, environmentalists and environmentally minded citizens have succeeded in setting aside land from development to save these creatures. However, when a firestorm breaks out in these areas, the focus is usually on saving structures that border them and these habitats can be written off. In many cases, once they are gone, they are gone.

The threat of massive wildfires is real and growing. This book shows how a regional fire disaster came into being and explores the political, social, and personal aspects of that tragedy. My hope is that people will gain a greater awareness of the wildfire potential in their own areas and rise up to do something about it before it is too late. The time to learn from past mistakes is now.

HELL
ON
EARTH

Prologue

Jason Roberts* loved the San Bernardino Mountains, spending every available weekend enjoying them from the sleepy former logging town of Crestline, just fourteen miles up the highway from San Bernardino, to the touristy resort town at Big Bear Lake twenty-six miles farther east from Crestline. This day he had Lake Arrowhead and its clique of wakeboarding enthusiasts in his sights. The athletic college freshman also had a certain gorgeous blonde, a member of the wakeboard club, in his weekend plans.

Strong shifting winds buffeted his red Toyota Celica as he pointed the car onto State

---

* Not his real name.

Highway 18 and began his ascent through the thick, dry chaparral up the mountain toward his long-anticipated day and a half away from his waiter duties at an overpriced theme restaurant in Rancho Cucamonga. In the restaurant business, any weekend day off was rare. Jason wouldn't have to report for work again until 5:00 P.M. Sunday, leaving plenty of time to get in some much-needed R & R and perhaps get lucky.

His driver's trance was broken when a gray van cut him off at the Old Waterman Canyon on-ramp just a few miles up the highway from the northern neighborhoods of San Bernardino, and he would later say he couldn't believe his eyes when he saw flaming objects being tossed from the speeding vehicle into the brush.

He grabbed his cell phone and dialed 911 as he struggled to keep up with the still-accelerating van. He caught a glimpse of the driver as the road and the van swung to the left, and then it was gone.

By this time Jason realized his emergency call had failed. The serpentine four-lane road boxed in by steep canyons and cliffs is a black hole for cellular phones, and in his frustration with his underpowered car and the elusive phone signal, Jason realized he had not gotten the license number of the van.

Meanwhile, at a home located on Skyland Drive in Crestline, overlooking all of Waterman Canyon, Clive Smith* was watching the same scene through binoculars. From his vantage point on the Rim, Smith called the authorities on his

---

* Not his real name.

cordless phone. The insistent beeping in his ear told him the battery of his phone was dying. He ran to the back bedroom and picked up the landline phone, nearly screaming into the receiver: "Some guys in a van are throwing fire into the brush!"

A few minutes later, at a pullout overlooking San Bernardino and the entire Inland Empire, Jason would finally reach the California Highway Patrol emergency center. "Some guys in a gray Chevy van are throwing Molotov cocktails out their windows!" he yelled. The operator logged the call at 0959 Saturday morning, October 25, 2003. Jason didn't know it, but at 9:16 a fire had already started in Old Waterman Canyon through thick stands of manzanita and other brush; by 9:35 it had spread to the San Bernardino city limits, and at 9:44—a quarter of an hour after Jason began his ascent of the mountains—officials had already closed State Highway 18 at the bottom of the hill. The Old Fire, as it would come to be known, had begun, only ten miles to the east of the Grand Prix Fire, which had already been raging out of control for days.

SALLY DESMOND° WAS TIRED. For nearly a week she had been providing support services for firefighters attacking the Grand Prix Fire, which ran from Fontana and Rialto to the west and the upscale bedroom community of Rancho Cucamonga to the east along an area, traversed by Interstates 10 and 15, that Southern Californians call the Foothill

---

° Not her real name.

Corridor. Desmond, a second-year student in the nationally recognized firefighter training program at Crafton Hills College in Yucaipa, east of San Bernardino, was called to the front line by the battalion chief.

"There's a fire in Waterman Canyon and it's out of control," he said, choking on his words through the smoke and ash. "Get over there and see what you can do. Check in with Captain Garrett."

The tall, studious girl inquired about the Waterman crew's needs in terms of drinking water, food, and first-aid support materials. "What do I need to grab?" she asked.

"Come again?"

The battalion chief was genuinely puzzled by her question.

"Will they need us to get them some chow to their front line, sir?"

The chief chuckled and shook his head. "Honey," he smiled, "you're going there to fight a fire. The only thing you'll need to grab when you get there is a hose."

Investigators already knew the blaze they had dubbed the Grand Prix Fire—named for Grand Prix Drive and Shetland Lane, in Northern Fontana near its point of origin—had been started by an arsonist. The firefighters were exhausted; they had been fighting the monstrous flames in hot, dry weather since October 21, a Tuesday. Now it was Saturday, and the Grand Prix Fire was threatening to burn north into the Cajon Pass, Devore, and Lytle Creek areas along Interstate 15, the route from the Inland Empire of Southern California to Las Vegas. Now they were forced to send rookies, such as Desmond, to provide support for the San Bernardino City and California Department of Forestry and Fire Protection (CDF) crews in Waterman Canyon.

The situation was about to get much worse.

ONE

December 2002

As 2002 was coming to an end, Peter Brierty was frustrated. As the fire marshal of San Bernardino County, the largest county in the United States, he was accustomed to having scant resources and being forced to beg for equipment and gear other fire departments took for granted. He didn't mind the fight; he usually got the board of supervisors or his immediate superiors to pony up the money, and unlike many in his position, he had a knack for knowing where the deepest government pockets were at any given time. This skill, coupled with his talent as a grant writer, would give the mountain communities a fighting chance against an enemy most didn't yet even realize existed.

Normally not one to remain down for long, the imposing 6-foot, 4-inch figure, who bears a strong resemblance to television's Dr. Phil, was fighting the only battle he truly feared: public apathy.

For nearly a year, Brierty and his core staff had done everything in their power to get the word out to the residents and property owners across the densely forested and over-populated San Bernardino Mountains. Tonight he would lead a public meeting in the resort community of Lake Arrowhead concerning the bark beetle problem and the drought, which were coming together to bring a looming danger and the potential for unprecedented destruction to these mountains.

His mission was to call the town to action and make them care about the problem. He knew that if Lake Arrowhead residents failed to follow through once he laid all his cards on the table, there would be no hope. Soon there would be a forest fire unlike any in recorded history, and likely many of the people he would address that evening would perish.

At 7:00 P.M. on Tuesday, December 3, 2002, Fire Marshal Brierty began his presentation. His earlier fears proved founded—he was preaching to the converted, addressing exactly fifty-two audience members, twenty or so who would have attended anyway as members of the local fire safe council. Still, he forged ahead, hoping those who came would spread the word and there would be an accurate report the following Thursday in the *Mountain News*, the local weekly. None of the other invited media in the area had chosen to report this tragedy in the making.

"I'm a bit disappointed," he began, "not because you're all here; I can't express enough my thanks to you for coming tonight. I'm disappointed because those who really stand to lose the most, all your friends and neighbors, aren't here to learn what you and I already know: This entire mountain, from Cedarpines Park and Crestline in the west to Big Bear in the east, is an unmitigated disaster waiting to happen."

Brierty did his best and it was duly reported in Thursday's paper, but he still felt as though the entire exercise was a waste of time. The following Monday he drafted a report and sent it off to his boss, San Bernardino County Fire Chief Peter Hills.

Almost two weeks later, when Chief Hills finally had a moment to digest Brierty's report, he took his colleague to lunch. "Pete, there's an old saying about leading horses to water," he said. "You've done your best."

Brierty was still frustrated. "If we don't do something, I mean something of substance, who do you think they'll blame when the firestorm comes?" Brierty asked, a tinge of sarcasm mixed with sadness in his voice.

During lunch Brierty proposed a notice be sent to residents to abate the dead and dying trees, which could become the fuel for the fire he so dreaded. While Hills thought the proposal had merit, he also knew it would be a tough sell. Many of the dead trees still looked healthy and green, and both men knew hitting moneyed people in their wallets for unexpected tree removal costs at their vacation homes would not be well received.

The meeting ended on some other business the two needed to discuss before conversation would turn to their families and the coming holidays. With the weather in a cooling trend, complete with rainy days and snow flurries, talk of the forthcoming tree removal orders could wait until after the New Year. Besides, Brierty thought, there would be no way he would ever be denied support from the governor's office once Sacramento realized the extent of the danger. Surely Governor Joseph Graham "Gray" Davis would have no argument with declaring the region a disaster area and open the tap for state and federal funding to help pay for the removal of some thirty thousand dead trees.

Brierty couldn't have been more wrong.

IN AN OFFICE TUCKED behind a sprawling industrial complex off Interstate 5 near Valencia, California, Ron Wieregard,* a thirty-year veteran of the U.S. Forest Service Arson Investigation Detail, received a call on a seldom-used line to which only a handful of upper-echelon state and county officials were privy. He jumped for the old-style Western Electric government-issue telephone with the overly loud ring.

"Arson Investigation, this is Wieregard." His heart pounded whenever this phone rang.

"Ron," said the caller, trying to mask the combination of fear and anger he felt, "we've got a problem." The station chief would only call Wieregard as a last resort, something the veteran arson cop knew. In the fraternal patter of interagency rivalry, the fire guys let it be known they maintained their own staff of arson pros. They would only go outside their own ranks if they were completely vexed and the general public was at risk.

---

* Not his real name.

Within an hour Wieregard and his chief investigator, Allison Murray, were at the scene of an arson call at a Dumpster behind a group of town houses.

The station chief took the pair to an alley two hundred yards away from the Dumpster crime scene. "What do you smell?" the chief asked.

Murray was the first to catch the distinctive whiff of gunpowder, the kind of smell found at any local service organization's retail fireworks stand on a hot July day.

"Man," Wieregard chimed in, "that's strong, huh?"

Their hope, at this point, was to garner some correlative evidence from a series of Dumpster fires that had been plaguing the area. In recent weeks the investigators had come to the definite conclusion there was a serial arsonist at work. The same incendiary devices were found at each location, and by now there had been so many fires started that a pattern was beginning to emerge. The fires in Valencia represented the arsonist's most northern work area, south along Interstate 5, along the Interstate 210 through the San Gabriel Valley, and ending east along Interstate 10 in Fontana.

Wieregard walked away from the group and found a quiet corner. It was never easy to call home and ask his wife to not wait up. It would be another long night. As he wished her good night and made his way to the hood of the car that would double for a command post of sorts, he knew she would be up and waiting when he came home.

TWO

An Arsonist's World

Oftentimes, when a writer works on a project like this, there is a large and necessary learning curve required in order to do justice to the subject matter. The firefighting profession is a highly technical one, life-and-death decisions are often made in an instant, and everyone engaged in fighting a fire must be working off the same page at all times. The firefighting professionals were so proud, so open, and so excited to share their experiences and vast knowledge for the benefit of the readers of this book.

However, to get the whole story, I also had to learn the influences of drought and global warming on wildfire by researching the latest literature from the world's experts in the scientific community. And finally, it was also necessary to learn every-thing possible about the human perpetrator—the arsonist. That portion of my research was anything but an enjoyable pursuit and, in fact, was horrifying.

One weekend day as rain poured down in sheets the ac-quaintance of the enemy was made.

This happened quite by accident as I took a break from writing and hopped onto the Internet for some fact checking. Within an hour an alternative Internet newsgroup was dis-covered, the primary purpose of which is to discuss arson and arsonists.

I made a post that included my phone number, and a cou-ple of hours later the phone rang. A young-sounding voice greeted me. In my opinion, either he was not too bright or he must have possessed some secret desire to be caught. Or he was so consumed by momentary ego-gratification that he didn't stop to consider that I might have been a member of law enforcement or an arson investigator.

Based on what he would eventually tell me, I pegged his age at late twenties or early thirties. What he would reveal so casually to me in that brief phone call would shake me to the core.

It would be easy to dismiss the caller's words and de-meanor as so much braggadocio, but I believed him. And while I am certain many use these newsgroups as a learning tool to prevent arson-related fires, what reason would a young man have to surf arson- and fire-related newsgroups unless he, at the very least, had some propensity for the subject? If my caller indeed wasn't an arsonist, he was certainly enam-ored of arson. In any event, his words rang true.

He wouldn't give me a name. Here is what he had to say.

"You don't know me, but you do. I'm your son's best friend, and I go to your church. I hate you. More like, I hate me, but I can't admit that."

He sounded a bit too polished in his pitch. I listened skeptically, believing or perhaps hoping the call was someone's bad idea of a practical joke.

"Something happened to me," he blithely continued, "to make my world way different than yours will ever be. I mask what I am so well, sometimes I go for months or years before I have to do it again.

"I started when I was seven.

"My first boners happened whenever I went to the playground at the school behind our house and struck matches on the cement. I would get hypnotized just watching the flames. Sometimes piles of leaves, sometimes paper and boxes. When I was nine I burned my first Dumpster. It was behind McDonald's.

"When I was thirteen I stole that movie *Backdraft* from a video store at the mall. I didn't want to ask my mom to rent it because I didn't want anyone to know about my love for flames, my love for fire. Burn, motherfucker, burn, goddamn it! There's a song that says that. I got hard, real hard, one night when I was fourteen and my mom was out on a date with the asshole she ended up getting married to.

"So I put on *Backdraft* and I whipped it out. I got so fucking horny; I was just beating to the flames, and I started a fire right there in my fuckin' room, ya know?

"So, anyways, I finish up just when the smoke alarm goes

off. I got turned on all over again. Then I just put it out with bathroom water and slept like a fuckin' baby.

"I do what I want, you know? I feel like starting [a fire], I just find something I want to burn and it's gone. I burned some houses, ya know? Like, no one was living in them, or nothing. They was still being built, so me and my friend Tony, who gets off on fire like me, well, we torched these houses. It was on the news and everything. It was so fuckin' cool, ya know?

"We rode our bikes down after the fire got going and we watched them put it out. I thought we were busted because we were watching it all go up and all of a sudden behind us we hear this really loud voice. He says, 'Hey, guys, can I talk to you?' It was a cop. So we walk over to the curb and he says, 'You'll have to get out of here. You can't be here while the firemen are working.'

" 'It's cool,' we said. And we rode over to Tony's grandma's mobile home park and started a fire in the Dumpster behind the clubhouse."

PERHAPS JIM KOURI, THE former director of campus safety at Saint Peter's College in New Jersey and the fifth vice president for the fourteen-thousand-member National Association of Chiefs of Police, describes it best when he says the American criminal-justice establishment views arson as one of the most heinous and aggravated offenses, "for it not only endangers human life and the security of home, but it also displays the highest degree of moral recklessness and depravity in the perpetrator." The example just cited proves the point.

Taken to the extreme, arson is a crime many in law

enforcement—including Kouri—believe deserves more severe penalties, including death.

"Not only does arson take lives, but it is estimated that arsonists have an incredible impact on our nation's economy," says Kouri. "The cost of insured fire loss within the United States exceeds thirteen billion dollars annually. More important, over fourteen thousand lives are lost in fires each year. And arson is known as one of the more successful criminal endeavors in the nation."

Kouri emphasizes the arson investigator's position is unique in that the evidence that a crime has been committed is not usually visible and, when it is, it must be preserved until investigators arrive. Too often, evidence of the crime of arson goes up in smoke along with everything else in the blaze. Whether working in the public sector for a police or fire department or in the private sector for a private detective agency or insurance company, the arson investigator must be qualified to give expert testimony in a court of law as to his or her opinion that the fire was deliberately set by human hands.

As in all crimes, the usual information must be compiled at the crime scene and the supporting evidence must be preserved and protected. The unique circumstances surrounding the crime of arson are responsible for the dubious distinction of being a successful criminal act.

"Unlike the crimes of burglary, robbery, and other offenses, the motives of the arsonist are hidden and difficult to identify," Kouri says. "There are a number of reasons why people start fires. Unfortunately, the only time a crime pattern may be used to arrest an arsonist is when it is the work of a pyromaniac or 'firebug,' or a professional 'torch,' who is hired because of his skill in disguising arson, usually for insurance fraud purposes."

Kouri describes the pyromaniac as a mentally ill person who usually operates in the same general geographic area of the community and who nearly always employs the same fire-starting method. Kouri says the fire-starting method can be used to associate the firebug with more than one case of arson.

INCENDIARISM

The arsonist is not deterred by the occupancy of a dwelling or the dangers faced by others in fighting a fire. In fact, the presence of occupants, in some instances, may actually provide the motive for the crime. Yet in spite of the great hazard and the financial losses resulting from this crime, the majority of arson fires go undetected.

The Fraud and Arson Bureau of the American Insurance Association claims that a great many fires are indeed arson and that the number of incendiary or suspicious fires has tripled since 1956. These figures reveal that there were over sixty-five thousand such fires with an average cost of fifty thousand dollars per fire. These figures do not include incendiarism involving aircraft, boats, motor vehicles, forests, et cetera.

The unique qualities of the incendiary fire affect, to varying degrees, the investigative efforts by detectives or fire marshals to prove that a crime was committed.

One of the more common obstructions investigators encounter is the absence of eyewitness testimony to the commission of the crime. The arsonist usually uses an ignition device that allows ample time to flee the scene before the fire is discovered, or his or her act is committed at night, when the chance of being apprehended is reduced. Also, the fire by collapsing walls and the disturbance of the crime scene by firefighters will generally destroy most of the criminal's traces or obliterate them.

"This all means that the investigator must conduct a determined search at the scene and seek out the assistance provided by scientific aids to uncover any physical traces left by the arsonist," Kouri says. "Inquiries must always extend into the backgrounds of persons directly or indirectly affected by the fire and who may be likely to have a motive for committing the act."

Kouri believes that, like other investigations, these efforts are necessary to establish the commission of a crime and the identification of suspects.

Unlike most criminal investigations, however, in arson the detective's initial tasks are to positively determine and be able to prove that a crime was actually committed. If this cannot be established, then there is no criminal case and the fire will be classified as accidental in nature. Two points that must be proven are that burning occurred and that the fire was willfully ignited by a person or persons.

FIRE KILLS IN MANY WAYS

On March 13, 2006, firefighters in the Texas Panhandle came upon a scene that showed the horrors of wildland fires. In a ravine they found a charred vehicle. Nearby were four burned bodies. Investigators later speculated that the four oil workers had become disoriented in the smoke, driven their car into a ravine, and then finally made a desperate attempt to run for their lives before being outrun by flames. At least four other people died running from the flames in the Texas fires. That was also the fate of as many as ten thousand head of cattle, many of which were found pinned against fences and burned alive because they could not run farther.[1]

The fires killed in other ways also. On the day

IDENTIFYING THE PERPETRATOR

After the investigator successfully establishes the commission of a crime, the detectives or fire marshals assigned to the case must introduce evidence to show the suspect committed the act of arson with criminal intent.

"Intent may be proven by showing the interior of the building was in some manner altered to either accelerate the burning rate (venting provided by breaking windows or making other openings) or delay the discovery of the fire by locking doors, or by covering windows," Kouri says.

Criminal intent may also be established by showing the

before the four bodies were found, four people were killed in a nine-car pileup near Groom, Texas, after heavy smoke from grass fires and blowing dirt dropped visibility to zero on a rural highway.[2]

We often hear that many more people die from smoke and carbon monoxide inhalation than from actually being burned by fires. This is true, but many people believe this only happens in structure fires, in which people are trapped indoors in enclosed spaces. Think again. In the widespread 1987 wildfires in the People's Republic of China, 221 people were killed, most of them not by flames but because of high concentrations of carbon monoxide in their villages.[3]

Sources: 1. Associated Press news reports, March 14, 2006. 2. Ibid. 3. Johann G. Goldammer, *Our Planet,* vol. 9 (New York: United Nations Environment Programme, 1998).

occupant failed to turn on an alarm or to extinguish a blaze when he had the opportunity to do so. Detectives are deeply suspicious if the alarm system is not operating or there appears to have been tampering with the sprinkler system and standpipe.

PROFILE OF THE ARSONIST AND MOTIVES FOR ARSON

The general profile of the arsonist is a white male under the age of twenty-one years, with a middle- to upper-middle-class background and above average intelligence.

One survey found that 65.2 percent of arsonists arrested for their crimes are under the age of eighteen. Fires of an incendiary nature are often set by pyromaniacs but are also ignited by people seeking economic gain or attempting to conceal another crime such as a homicide. These fires may also be started for the purposes of revenge, intimidation, and extortion.

There are many motives for arson. For example, a fire started for the purpose of fraud can be meant to destroy property in order to collect insurance money, or as a means of preventing serious financial loss, or to terminate an unprofitable lease and for other reasons that bring the arsonist or his employer financial gain.

The arsonist may be the proprietor or operator of a business. He or she may be a person totally unrelated to the fire victim, such as an insurance adjuster or contractor seeking additional work. The arsonist may also be an investor in an unprofitable business who is neither directly involved in the operation nor a relative of the business's principals. Also, there have been an alarming number of fires set by firefighters or security officers who wish to be recognized for their heroism in fighting the blaze they started.

Detectives or officers investigating the cause of a commercial fire are also on the alert for indications of excessive fire insurance coverage and whether the premium payments are soon due. Inquiries are made to determine whether the business was on sound financial ground and whether the operator or anyone connected with management and administration of the business has a history of business ventures destroyed or damaged by fires.

"One common fraudulent practice is to insure a store's stock actually in existence at the time the policy is obtained," Kouri says. "After the insurance is secured, the store's inventory is reduced either by conducting a bogus sale or by removing the merchandise to another location."

After the fire, a claim is submitted to cover the cost of the stock that was allegedly lost in the flames. The businessperson then sells the "bait" now damaged by smoke, fire, or water to the insurance company as adjustment against the loss.

ARSON: A MEANS OF INTIMIDATION

Arson is also a tool of the extortionist and racketeer of various sorts—labor and organized crime—or simply used by one businessperson against another as a means of obtaining demands. Usually, it is difficult to solve cases of arson resulting from labor disputes or from attempts of a syndicate-controlled business to expand into other commercial enterprises. These fires are the work of professional criminals who may not even be local residents. During the early days of the Teamsters Union, when the legendary Jimmy Hoffa began his rise to the top of organized labor, there were several fires allegedly ignited by militant members of the labor movement. In fact, according to experts on Hoffa and the Teamsters, Hoffa took part in arson and on one occasion watched a friend and mentor burn from a fire that was improperly ignited with gasoline.

Other examples of arson for intimidation are the mysterious

residential fires or bombings that occur in communities experiencing racial or ethnic conflicts. Flames engulfing a home in the dead of night can carry a strong message that will tend to curb the liberal zeal of a member of the majority group, just as they can discourage and frighten members of a minority group from "moving into" a previously restricted neighborhood.

THE REVENGE FIRE

The revenge fire generally occurs at night, particularly when the object of attack is someone's home. There may be a number of reasons for a person wanting to burn the property of another. Some are petty or maybe even imagined, while others have greater significance. Jealousy, employee dissatisfaction, the anger of a competitor, the pain of a jilted lover, and a tenant-landlord disagreement are all characteristic motives of the revenge fire.

A classroom or schoolhouse is usually ignited by a juvenile arsonist in retaliation for a real or imagined wrong such as reprimand, a suspension from school, or failing grades.

There have been several cases of volunteer firefighters or private security officers starting fires they later help to extinguish. Even private citizens may start fires for publicity. It is not uncommon to read a newspaper story reporting the heroic efforts of a citizen who, upon seeing a fire, rushed into an apartment building and saved its residents, and later it is discovered that the "heroic" citizen is actually the person who started the blaze.

PERSONAL GRATIFICATION

As is the case with the arsonist described earlier in this chapter, a large number of fires are started by people suffering from mental or emotional illness. Many of these are pyromaniacs who set fires for personal pleasure without realizing, comprehending, or caring about the gravity of their acts.

And also like my interview subject, many pyromaniacs are known to derive sexual pleasure from their fires.

Generally, the pyromaniac has no rational reason for the behavior. The fire is started on impulse. The first attempts are directed at small structures—fences, trash cans, and outbuildings. Later, the pyromaniac graduates to starting fires much larger in nature, in places such as homes, apartment houses, factories, and storefronts.

Having started the fire, the pyromaniac often remains at the scene to make certain the fire is well ignited and then mingles with the crowd of bystanders. Clever investigators will always photograph the crowds surrounding the fire scene.

A series of fires occurring under similar circumstances identifies the work of a pyromaniac or serial arsonist. The crime pattern and the pyromaniac's delight in watching a fire are the usual elements that lead to the firebug's arrest.

"I GOT A PLAN," my interview subject said as the phone call drew to a close. "I'm gonna start the biggest fire ever. It'll be beautiful. I'm getting a wood just thinking about it."

I asked him if he ever sought help for his problem or if he even recognized it as such.

"I had to go to some dumb-ass counseling when I was about seventeen or eighteen years old. They asked in the paperwork if I like starting fires, or some bullshit like that. So I said no, like I would tell them the truth in the first place.

"My mom made me go because she said I was all fucked up and liked Marilyn Manson and because I told her I didn't believe in Christ as my Savior. Fuck that shit. Give me Satan any day. He likes fire, too.

"I don't know when I first started liking fire. All I know is that I do." And he then rang off with a casual "Later!"

THREE

Please, Governor

In the spring of 2003 the last twenty-two lookout towers maintained by the State of California finally fell victim to the state's budget shortfall, according to an Associated Press story in *The Sacramento Bee* that said the CDF wouldn't staff the summer lookouts—a move that was intended to save seven hundred thousand dollars.

When the budget ax fell, residents and officials near the lookouts began lobbying lawmakers to keep them open. In fact, in the prior year, 2002, California governor Davis found emergency money to temporarily open ten additional lookouts during what was projected to be an extreme fire season.

While only 7 percent of fires were first spotted from the state lookouts in 2001, that still represented a total of 415 fires that possibly would have taken hold and done greater damage had the lookouts not been staffed.

The Forest Service still maintains about sixty of the roughly six hundred lookouts it operated in the years after World War II and is reluctant to abandon this treasured relic of its past.

The state's remaining lookouts are all in Northern California. The state once operated as many as seventy-seven lookouts statewide, but abandoned those in Southern California years ago after the region's smog and haze made it more difficult to spot fires.

Besides closing the lookouts, the CDF also consolidated two air tanker bases to save another seven hundred thousand dollars, including closing the forty-two-year-old Ukiah Air Attack Base and reassigning its two tankers to bases in Chico and Santa Rosa. The state's reasoning was that with faster and bigger air tankers in those two bases, firefighters could still make a twenty-minute arrival time to any location in the state.

"This is a low-down dirty shame," a career air tanker pilot, Jim Barnes, said. As reported in the *Wildfire News* after the firestorms, Barnes, a veteran flier with thousands of flight deck hours spent on firefighting aircraft, believes the launching and staging zone was as important as the actual aircraft fighting the fire.

"Ukiah Air Attack Base played a vital role in delivering rapid response, initial attack, and aerial fire suppression for its entire zone of influence." Barnes said the zone of influence includes all the territory within a twenty-minute radius of Ukiah Airport, including roughly 1,776 square miles of what is called wildland/urban interface and forest area.

Barnes disagreed with the CDF assessment that the fire protection gap left by Ukiah's closing could be filled from other air attack bases. For starters, Ukiah has the benefit of being generally fog-free during the summer.

"On our hottest days, rising air in the interior of California causes costal fog to be sucked into low coastal areas like Santa Rosa and Rohnerville," the airman said. "Shrouded by dense morning fog, the bases are rendered useless until at least 11:00 A.M. or later." It is also well known that because of the generally higher temperatures in the Ukiah area, the Ukiah Airport will remain fog-free while Rohnerville and Santa Rosa are socked in with pea-soup fog. Therefore, there are a number of days during fire season that Rohnerville is open only a couple of hours a day.

Barnes points out that Sonoma County Airport is tower controlled, with a high level of nonfirefighting aircraft traffic, and says firefighting aircraft can suffer significant delays there during periods of peak activity.

"With Ukiah open," he said, "we simply elected to load up there if the fire was in the Lake or Mendocino areas. Also, Chico Airport is at the edge of the twenty-minute response from Ukiah. Turnarounds that took ten minutes out of Ukiah could now take thirty to forty-five minutes out of Chico, depending on the fire's location.

"If we lose just one fire as a result of closing Ukiah Air Attack Base, the cost may outweigh staffing Ukiah for many, many years," Barnes declared. He hoped the issue would be revisited, rather than someday being cited in an investigation as a "contributing factor in failing to stop a fire in the initial attack stage, thereby resulting in catastrophic losses."

MEANWHILE, AS THE STATE grappled with budget issues for the CDF and other agencies, the dead zone of trees lost to the drought and bark beetle infestation was expanded from 66,000 to 80,000 acres. Predictions that the area would reach an unprecedented 150,000 to 170,000 acres by year's end were just more harbingers of a fierce fire season.

While news reports would stubbornly indicate there were only about 150,000 dead trees, the more realistic estimate was 320,000, calculated by simply using a formula: an average of four dead trees per acre. Inspections revealed some areas already had a 30 to 40 percent tree mortality rate, while others showed tree deaths of 80 to 100 percent in some areas.

Royce E. "Rocky" Saunders, Governor Davis's Office of Emergency Services representative, repeatedly stated that he would not recommend his boss issue a Declaration of Emergency. In fact, Saunders continued to assail anyone who even attempted to suggest the governor issue a Declaration of Emergency, even when it was clearly stated no appropriation funding would be attached.

Saunders, based out of Sacramento, was the branch chief of disaster assistance programs for the Office of Emergency Services. Local fire safe councils strongly believed a Declaration of Emergency, even without any funding attached, would empower the CDF to expand their role in removing dead trees on private land areas.

"The CDF would be tasked to remove dead trees without regard to active beetle infestation, and would not be limited by private property issues," CDF Deputy Chief Dave Neff revealed. "Immediate action toward fuel reduction would assist other agencies, alleviate critical fire fuels in private land areas, and reduce the potential of fire spread." Savings to the state through mitigation instead of fire suppression is conservatively estimated to be seventeen to twenty dollars for each dollar spent by the CDF in an expanded role.

Saunders later advised a tricounty audience in Temecula, attended by fire officials, fire safe council members, and government representatives, that he would not discuss such a proposal, nor recommend it to the governor. Temecula is a rapidly growing, mostly residential city located in western Riverside County just north of San Diego County. The area is surrounded by dense brush and is plagued yearly by wildfire.

A state employee who wished to remain anonymous due to his fear of "whistleblower retaliation" said, "Davis's lack of action stood in the way of achieving major assistance programs, and hindered in bringing the disaster to the attention of FEMA."

To this day, fire officials agree that forest conditions had never been worse. Each time the winds picked up, local fire agencies prepared for the imminent fire and the resulting disaster.

No clearer evidence of a forest in crisis could have been demonstrated than by the number of trees that fell onto Lake Arrowhead area homes. The trees also began toppling onto roads and utility lines at an alarming rate. Because of the danger of trees falling on power lines and sparking fires,

Southern California Edison officials began contemplating a plan to completely cut off electricity in the San Bernardino Mountains if a certain combination of wind conditions and humidity levels was reached.

One day a tree fell onto a Crest Estates home on Ridgecrest Drive in Lake Arrowhead. The fallen tree revealed a root ball slightly larger than the diameter of the trunk, and the treetop had broken across the upper railing of the patio deck, finally coming to rest on the lower deck railing.

Wood moisture was nonexistent in the tree and, upon inspection, conditions indicated advanced interior decay. This insect-laden white fir was a classic example of drought-related mortality causing an acute forest disaster situation and high fire danger.

But despite the absence of state funding assistance, agencies continued to seek solutions to the mountain ecological disaster.

U.S. Forest Service Mountaintop District Ranger Allison Stewart and Bob Sommer, a USFS battalion chief, announced plans to create a six-hundred-foot shaded fuel break along the boundary of the national forest adjacent to private lands (a shaded fuel break is one that still contains some brush and trees). Stimulated by Congressman Jerry Lewis, USFS personnel took emergency steps to reduce fire fuels around the San Bernardino Mountains communities.

But in the absence of financial assistance from state and federal agencies, the innovative solutions and good ideas were difficult to put into practice.

It would be difficult enough to get FEMA funding to deal with an impending disaster rather than one that had already happened, but without a Declaration of Emergency from the governor, access to FEMA funding was all but impossible. There was, however, someone in Washington who was listening.

Congressman Jerry Lewis, a senior member of the House Appropriations Committee, had been attempting for months to reallocate $3.3 million of unspent disaster mitigation funds as a way of providing some much-needed fire danger reduction through tree removal on private lands near the national forest.

"Property owners in the San Bernardino Mountains were surrounded by dead and dying trees and faced either losing their homes to catastrophic fire or losing them because of the prohibitive cost of removing those trees," Lewis recalled.

The funds requested for the bark beetle problem would be allocated from an unused portion of grants secured by Lewis in 1999 and 2000 for seismic retrofitting for buildings at the California State University–San Bernardino (CSUSB). The university completed that project and was no longer in need of the leftover money.

Although the funds remained unspent, FEMA officials told Lewis they needed specific language from Congress directing the money to be transferred for use on the bark beetle problem. Because of the high cost of tree removal in many locations, Lewis personally urged Deputy FEMA Director Mike Brown to provide a mitigation grant to state and local authorities to use on private property.*

---

* Mike Brown would later be promoted to FEMA director during the George W. Bush administration and would become the focus of much of the public's ire at the handling of disaster efforts connected to Hurricane Katrina in 2005.

To resolve the legal limitations on disaster funds earmarked for use, Lewis agreed to introduce legislation directing the reallocation of funds from the CSUSB project.

As a senior member of the House Appropriations Committee, Lewis was well versed in making personal contacts with those who have the power to push a spending bill through the maze of governmental red tape.

At first Lewis believed it was possible to ask FEMA for a simple transfer of funds, but it soon became clear FEMA was not going to do it without legislation. Lewis personally pushed the language through and made numerous eleventh-hour phone calls to other congressional leaders reminding them of the importance of the bill. Public officials responsible for the San Bernardino Mountains communities knew $3.3 million was a drop in the bucket compared to the magnitude of the problem, but it was a start, and the money would help many mountain home owners.

The money was distributed to agencies, such as the CDF, that used the funds to remove dead trees in badly affected areas. It was enough to get the ball rolling, and Lewis didn't stop there. He committed himself to finding other emergency funding sources as well.

One plan involved the Bush administration's Healthy Forest Initiative.

The congressman heavily lobbied the administration to consider using the San Bernardino Mountains as the pilot example of removal and reforestation through the Healthy Forest plan. Lewis visited the area with then secretary of agriculture Ann M. Veneman, who described the unimaginable scene to the president.

Meanwhile, it was clear to everyone in the governor's camp that Davis was suffering a major image problem. He appeared completely ineffective.

Polls indicated he was at the bottom of the barrel, and he would soon find himself in the midst of a major recall campaign.

The lion's share of the blame for California's energy costs crisis fell heavily on Davis's shoulders. And an unpopular "car tax," a proposed hike in vehicle registration fees, was showing signs of failing completely. There was a bipartisan hatred of Davis growing more each day, and he must have felt he had to do something, anything, to gain favor and turn his plummeting poll numbers around.

On March 7, 2003, Governor Gray Davis gave it his best shot when he signed the State of Emergency Declaration the firefighters and the affected mountain residents had wanted for so long.

The declaration went a long way toward cutting the red tape that had hampered local safety officials for nearly a year, as the action allowed the state and landowners to take steps to prevent a catastrophic fire in areas with dead and dying trees within Riverside, San Bernardino, and San Diego counties.

When he signed the declaration at the editorial offices of the Riverside *Press-Enterprise,* Davis said, "Trees on more than one hundred and fifty thousand acres have died and an estimated seventy-five thousand residents are threatened by catastrophic wildfire, injury, and property damage from falling trees. My action provides landowners with the regulatory relief necessary to quickly remove dead and dying trees from their property."

Davis directed the CDF to take immediate steps to protect public safety by clearing effective routes for evacuation and emergency response and by establishing safe fire evacuation centers. Ever the career politician, Davis continued, "This is a clear disaster, wreaking havoc with trees in the inland counties. We need to get on it as quickly as possible." Many suspected the governor's change of heart and call to action was simply a last-ditch effort to turn around plummeting poll numbers. They wondered why the "clear disaster" only became clear to Governor Davis once his political career was in jeopardy.

The declaration also suspended both the requirements for the prior notification for emergency timber removal and the limitation on the amount of dead, dying, or diseased trees that can be removed. Davis's action also allowed licensed tree removal contractors, as well as licensed timber operators, to be hired by landowners to remove affected trees. Collectively, his actions should have provided a larger pool of contractors for landowners to hire from and expedited removal of flammable vegetation. Many just believed it was a case of too little too late.

In the days following the governor's declaration, a capacity crowd, only now searching for solid answers to the mountains' tree problem, greeted a panel of fire and safety officials gathered on the dais in the main ballroom of the Lake Arrowhead Resort. In contrast to the sparsely attended public meeting with fire officials back in December, this time there was standing room only.

The town hall meeting was presented by the San Bernardino Mountains Fire Safe Council and covered a myriad of issues, including forest risk assessment, fuel removal programs, options for funding fire fuel removal, tax relief from the IRS, private funding, and insurance. Finally the officials had the public's attention.

Fire Marshal Brierty and David Caine of the fire safe council, who was the newly appointed president of the Arrowhead Communities Task Force, stepped to the podium and reported on the progress made to date in the seemingly endless war against the bark beetle and the dead trees.

"Since our first meeting, the governor of California has issued a State of Emergency Declaration. This is more than we had before, and it should cut a lot of the red tape that has hampered us so far," Brierty said. "That's the good news. The bad news is that no government agency is going to send in the treecutting cavalry to cut your trees down. That is still your responsibility."

The declaration issued by Davis removed many of the governmental procedures usually called for when mass tree removal is required. Normally a timber harvest plan would have preceded the downing of trees on private property when conducted by a licensed timber operator (LTO).

Also, by temporarily suspending certain state licensing requirements the governor's proclamation eased the way for LTOs from around the state to remove trees on private property.

Many property owners recall now how frustrated they were by the county's expectation that they remove their dead and bark beetle–infested trees within a thirty-day period. Just weeks before, Brierty had issued a number of tree removal citations and the people were not happy.

"I've got a bunch of dead trees that have to be removed, and I want to do it," said one audience member that night, addressing Fire Marshal Brierty. "But I can't get a tree remover to return my calls, and the one that did said the soonest he could get to me is three months. What am I supposed to do with a citation that gives me a month?"

Brierty answered, "Send me a letter to that effect and as long as we can see you are addressing the problem, we'll put you at the bottom of the pile." Brierty said he understood at the time the problem home owners were facing and assured them as long as they took action he wouldn't have to.

Brierty concluded with what would become the most prescient statement of the evening: "Now is the time to get out your pens and write the president and other federal officials. We want to make the governor's declaration as relevant as we can, so we need to make our voices heard in Washington. And we need to do it now."

The disaster was already much worse than the numbers Davis had quoted when he signed the declaration. A little over two months after the declaration was made, when Davis finally toured the devastated area, officials were conceding that at least 415,000 acres in the three counties were infested by the bark beetle, or about 650 square miles, an area a little under half the size of Rhode Island.

FOUR

The Bridge Fire

The first taste of what was to come began on Friday afternoon, September 5, 2003, when the "Bridge Fire" broke out in the San Bernardino National Forest near the City Creek ranger station, just north of Highland and northeast of the city of San Bernardino. Firefighters wondered if this was the Big One.

The look of the fire was made even more ominous by the meteorological conditions on that day. The initial cloud of smoke on the first day had merged into a towering thunderhead above, giving the impression of a cloud of smoke tens of thousands of feet high. This cloud could be seen from all over the mountains, the high desert, and the Inland Empire of Southern

California and caused concern over a wide area among people who were on edge, dreading what might come.

The fire immediately threatened Highland, population forty-five thousand, to the south and the five thousand residents of the mountain community of Running Springs, a few miles to the north and a few thousand feet higher.

By 6:00 P.M. on the following Tuesday the fire had blackened 1,352 acres, about 1.5 times the size of New York's Central Park, and had caused the evacuation of nearly fifteen hundred residents. It was burning in thick brush growing along steep canyon walls. When the decision was made to evacuate mountain residents in the Running Springs area, officials knew some people would go to the homes of family and friends and others would seek out lodging at local hotels, inns, and bed-and-breakfast facilities. But there were going to be other people who would need shelter until they could return to their homes.

Rim of the World High School above Lake Arrowhead (and overlooking the fire below to the southeast) was designated as the official Red Cross shelter for the Bridge Fire, and immediately it became a hub of activity as the facility was prepared for the influx of "temporary" occupants.

Red Cross volunteers brought in cots, bedding materials, food, and water and set up a registration area. Rim of the World School District personnel did everything from emptying trash cans to stocking restrooms and shower facilities. Still others briefed Red Cross volunteers on the location of light switches and extra supplies and helped prepare the campus for the displaced residents.

Tim Tipton, who lives on Spring Oak Drive in Running

Springs, said one of the first things he did was take his cats for boarding in nearby Blue Jay and then report to the high school with his wife, Sherry. "We made our decision to leave early on," Tipton remembered. "I was in the Air Force the last time I slept on a cot. We're staying elsewhere Saturday night because the metal cots squeak every time you move and we're not used to having small children around." But Tipton was very appreciative of the Red Cross's assistance that Friday night.

Debra Butler didn't arrive at the shelter from her Panorama Drive residence until Saturday morning. "A deputy came to my door and asked why I hadn't evacuated," Butler recalled. "I told him I didn't have transportation, plus I didn't have any place to take my pets."

Butler, who managed to get a few changes of clothing and several picture albums, noted that twenty-five minutes later a community service officer was at her home ready to take her to the high school. "And volunteers from the Humane Society were able to care for my animals—a dog, cat, and bird," Butler said as a small tear formed in the corner of her eye.

Bob Kastelic and his wife were in Las Vegas visiting friends and planned to take in a show Saturday night. "We decided to come back to find out firsthand what was going on with the fire." Kastelic informed a deputy stationed at the high school that he had a large gas can under the rear deck of his home. The officer promised the gas would be relocated to a safer storage area until the all clear was given.

Bob Mickelson was in Orange County when the fire broke out: "I called my wife, Jan, and told her to take the DeLorean and meet me off the hill. With the 330 closed, we decided to meet at the high school instead." The DeLorean, a rare stainless-steel sports car with gull-wing doors, was still at the residence on Pixie Drive that Saturday, but the Mickelsons did bring some paintings and jewelry with them to the evacuation center.

The Highway 330 closure also stranded several Running Springs residents at the bottom of the hill. One resident was long-haul truck driver Casey Moon: "I was on my way back from Utah and dropped my trailer off in Chino when I learned I wouldn't be able to get up Highway 330 to my home," he said. Moon, whose tractor was sitting at the Highway 30/330 Interchange at the bottom of the mountain in Highland, used his cell phone to call his wife.

"She had just been told to evacuate," Moon reported, "so I told her to come down Highway 18 and meet me here and we'd spend the night with our four-year-old son in the sleeper." However, the family eventually decided to get a motel room.

"During the Willow Fire several years ago," Moon said, "we had to evacuate and spend three days with my parents in another area of Running Springs."

Engine companies, tanker trucks, and bulldozers moved into position to battle the Bridge Fire along with aerial units stationed about ten miles from the fire at the former Norton Air Force Base in San Bernardino. "It's scary," Moon stated.

As the trucker was talking to reporters, a man and woman drove up inquiring if anyone needed a place to stay Friday night. While they didn't get any takers on their lodging offer, the dozen or so people at the Highland location were all impressed with their generosity.

As crews began cleanup duties along the highway, including the removal of fire retardant dropped from air tankers while fighting the fire over the weekend, Fire Marshal Brierty reported, "Everything's in a cooldown mode. In terms of any kind of a threat, the danger's passed."

Nevertheless, fire crews worked through the week monitoring hot spots and examining the cause of the fire. USFS investigators had already determined the fire was "man caused," even though they have yet to pinpoint the exact source of the flames.

"In an arson investigation, we start by ruling things out," said thirty-year veteran arson investigator Ron Huxman. "We've ruled out lightning, and we know the fire started close to the road, so that makes it suspicious."

Recalling the evacuation effort in Running Springs, Brierty said, "I was all over there on Friday, and there was absolutely no panic. While the various agencies are to be commended, I really believe the public was just as responsible for how well everything went." Brierty was especially impressed by the calm exhibited by the residents who were ordered to leave their homes.

"Our guys would tell the folks it was time to leave, and in calm, steady voices residents would say, 'We're ready to leave.' There were no honking horns or driving through yards," Brierty said. "The public did a tremendous job, and it went way better than almost everybody expected."

Mandatory evacuation was lifted at 6:00 P.M. Sunday, September 7, for Running Springs area residents, who were then allowed to return to their homes.

Throughout the entire weekend helicopter tankers, air

tankers, and firefighters from all over the state—some coming from as far north as San Jose and Stockton—joined forces to fight the Bridge Fire, the first major fire incident in the San Bernardino Mountains since Brierty's public meeting back in December, where he had spoken about the bark beetle infestation.

Brierty believes the orderly evacuation and the quick handling of the fire had much to do with preparedness. "If you were at the Tabletop Exercise at the resort three weeks ago, you heard, then, a lot of the scenarios which came to pass this weekend, and as you can see, being prepared made all the difference," he said. "That meeting had allowed safety officials from various public agencies to rehearse various mountain fire scenarios, including the evacuation of residents.

"To see everybody deploy in a calm but determined fashion was an incredible thing. For the last couple of years we have made the preplans—each community has devised a preplan and the chiefs each have a two-inch-thick three-ring binder showing safe zones and protection zones. This way, the thirty strike teams each had a copy of the fire protection plan."

Brierty was pleased by the reaction the out-of-area strike teams had to the preplan binders. "They looked at them and asked, 'How'd you make this so fast?'

"They looked at the color maps, descriptions, and these strike teams are saying they should do this in their own communities," Brierty said. "'We're taking these books home,' they said. And that was one of the most rewarding things that came out of this weekend for us. It was confirmation for us that we were dead-on."

Brierty also focused on the enormous public outreach officials and the fire safe councils had conducted in the previous months, citing as examples the many town hall meetings, block meetings, and volunteer activities.

"The fire safe councils have been in the loop from the beginning and they've helped out so much," Brierty said. "They were part of the planning process and now they are part of the success. To see the inner workings of the Bridge Fire event was very rewarding. Evacuations are difficult at best, but this couldn't have gone better."

Brierty said the questions and answers that were fielded at the numerous public information updates at Rim High School were in a positive vein. Speaking to one crowd of anxious residents assembled in the cavernous gymnasium, Brierty offered, "I think a lot of it is because of what you've done through the media, the public meetings, and having the public involved really paid off. This fire was a dress rehearsal for a show we hope we never see." He couldn't have known that just a scant two months later the biggest firestorms in recorded history would grip all of Southern California. In his coverage area alone there would be four major fires, and thousands of homes would vanish—charred memories guarded by ghostly chimneys were all that would remain after the flames had passed.

But for now, higher-ranking firefighters who attended the September 6 briefing at CDF headquarters at the old Norton Air Force Base were encouraged by the cool turn in the weather. Fortunately, winds that were expected to increase that Sunday and Monday never materialized. The heat and low humidity, however, would continue to be factors.

During the Saturday briefing at the Incident Command Post, officials said the Bridge Fire had consumed approximately 1,200 acres and burned a total of 1,350 acres before firefighters gained the upper hand late Sunday night.

Among those present at the Incident Command Post briefing were USFS Mountaintop District Ranger Allison Stewart, Running Springs Fire Chief Bill Smith, San Bernardino County Fire–Green Valley Lake Fire Chief Marvin Neville, former Running Springs fire chief Dan Wurl (now San Bernardino County Fire Department assistant chief), Pat Dennen (former Running Springs fire chief, now the San Bernardino County fire chief), and Mountain District Fire Chief Thom Wellman. That all-important briefing was attended by approximately seventy-five people.

Various firefighters presented the latest information. In addition to the obvious objectives of providing firefighter, structure, and public safety, the goals were to keep the fire north of the city of Highland and contain it from spreading farther into the mountain and toward communities there.

Firefighters were warned to "flag" and avoid rockfall danger zones, watch for rolling rocks on the fire line and the roads, have crews sleep only in designated camp areas, and watch for active fires or changes in "fire behavior."

Through the Mutual Aid system, firefighters and equipment came from all over Southern California, including units from Apple Valley in the Mojave Desert and the urban areas of Corona, Diamond Bar, and Los Angeles County. One of the largest contingencies came from the Pilot Rock, Prado, Oak Glen, and Los Angeles County CDF camps. A total of nineteen inmate hand crews were on scene from these camps.

The inmates could not become part of the hand crews without subsequent screening and approval. Once they passed their required skills tests they were put on a list and called out when needed. During the Bridge Fire, Camp Commander Lieutenant James Lewis oversaw the inmate crews.

A noon meeting was scheduled for the evacuation center at Rim High School on Sunday afternoon to update Running Springs area evacuees either staying at the center at Rim High or with friends or who were able to take advantage of a special "evacuees" rate offered at the Lake Arrowhead Resort.

Approximately two hundred people attended the meeting and wildly applauded fire representatives who reported that as long as fire conditions remained the same throughout the heat of the day, they could return to their homes by 6:00 P.M. Sunday, September 7, "but not before then, just in case things change on the fire lines," Chief Wellman said. A voluntary evacuation would still remain in effect. At that point the fire was 30 percent contained, twice the area of containment reported for the day before.

Twin Peaks Sheriff's Station Commander John Hernandez explained that while the roads were closed there would be forty-eight teams of law enforcement officers at checkpoints ensuring people did not enter restricted areas. He added that he had been meeting with Arrowbear Lake and Green Valley Lake representatives, keeping people in those nearby communities apprised of the fire situation.

Dr. Clint Harwick, superintendent of the Rim of the World Unified School District, announced a regular school schedule beginning on Monday, September 8. He added that he would station extra school buses and drivers at Charles Hoffman Elementary School in Running Springs just in case they were needed should the fire change direction.

"Our staff was outstanding," said Dr. Harwick. "The maintenance, operation, and transportation staffs as well as our custodians were wonderful. The high school administration was very helpful and I've received a lot of positive feedback on how the evacuation center at the high school was run. We're very grateful."

Red Cross volunteers logged six hundred hours helping mountain residents. The San Manuel tribe of the Serrano Indians provided hot meals on Saturday night, and Steve Sisk, Red Cross director of emergency services, said that everyone owed a lot to Dianne Strutt, who was the shelter manager. "She did a fabulous job," said Sisk, who added that over the weekend two hundred ninety people registered at the shelter.

Running Springs Fire Chief Bill Smith announced at the public meeting at Rim High that his department would receive federal firefighting funds to help pay for the costs associated with fighting the Bridge Fire.

Smith explained that at 3:00 A.M. that Saturday he received news that the U.S. Department of Homeland Security's Federal Emergency Management Agency (FEMA) had authorized the monies. The approval was made just hours after the California Office of Emergency Services contacted FEMA on behalf of the Running Springs Fire Department.

Applause rippled through the Rim High gymnasium when Chief Smith explained how the grant would help the department maximize firefighting efforts. The American Red Cross also had used the Rim High campus as an evacuation center.

"For a small department like ours," Smith continued, "a major fire like the Bridge Fire could totally break us. Now FEMA will pay seventy-five percent of the eligible firefighting and emergency response costs."

Smith indicated that the balance of the nonfederal cost share would be provided by county and local agencies. The affable fire chief's announcement was a relief to all in the gymnasium, because costs associated with the fire had already surpassed $1.3 million.

But the biggest round of applause of the day came when it was announced that nearly one thousand firefighters were working hard to protect property in the Running Springs, Smiley Park, Enchanted Forest, and Fredalba areas of the San Bernardino National Forest.

BUSH TO CONGRESS: FIX THE PROBLEM

The Bridge Fire, which, if conditions and preparations had been different, could have been the Big One, instantly became an important issue in the San Bernardino Mountains. It had served to erase any public apathy that might have remained.

Speaking in Redmond, Oregon—ironically as another major forest fire burned nearby—President Bush, accompanied by Secretary of Agriculture Ann Veneman (Veneman had visited Lake Arrowhead two months before the Bridge Fire and secured nearly $5 million in tree-cutting funds), described his plan. Bush spoke to hundreds of Oregon citizens, businesspeople, and logging industry workers and pulled no punches.

"The legislation makes forest health the priority, a high priority, when courts are forced to resolve disputes," Bush said. "And it places reasonable time limits on litigation after the public has had an opportunity to comment and a decision has been made. . . ."

Bush also added: "I've asked Congress to fix the problem. . . . Congress must move forward with this bill. It's a good, commonsense piece of legislation that will make our forests more healthy, that will protect old-growth stands, that will make it more likely endangered species will exist, that will protect our communities, that will make it easier for people to enjoy living on the edges of our national forests.

"The law, called the Healthy Forest Restoration Act, would bring government communities together to select high-priority projects relevant to local needs."

According to the official description of the initiative, "On May 23, 2002, Secretary Ann Veneman signed an historic agreement with seventeen western governors, county commissioners, state foresters, and tribal officials on a plan to make communities and the environment safer from wildfires through coordinating federal, state, and local action."

Bush began his speech that morning by relating how he had just toured "two fires that are burning in the area. It's hard to describe to our fellow citizens what it means to see a fire like we saw . . . big flames jumping from treetop to treetop, which reminds me about the brave men and women, what they have to face when they go in to fight the fires. . . ."

Bush continued, saying: "Ann [Veneman] was right. I was here a year ago. Unfortunately, when I came a year ago, I witnessed the effects of fires. I saw the Biscuit Fire and the

Squires Peak Fire. Both of them are devastating forest fires. They destroyed buildings and homes, changed lives, destroyed natural resources. . . ."

Bush asked for support for the initiative. "It seems like to me we ought to put a strategy in place to reduce the amount of money that we have to spend on emergency basis by managing our forests in a better, more commonsensical way," he said.

His statement was greeted by applause, but many firefighting professionals worried that the initiative wasn't as far-reaching as perhaps it could have been. Also, there were the usual concerns that the initiative was, in reality, a Republican gift to the lumber industry; that perhaps the initiative was a way for the federal government to circumvent environmental protection laws for the benefit of a very wealthy political support base.

THE BRIDGE FIRE: MOP-UP 100 PERCENT CONTAINED

The cause of the Bridge Fire was not immediately known. Forest Service Supervisor Gene Zimmerman said more such fires are likely for years to come because of the dry conditions in the area caused by persistent drought.

"We're primed for disaster," Zimmerman said. "It's taken a hundred years to get in this condition. It'll take us thirty years to get out of it."

As firefighters took the upper hand with the Bridge Fire, investigators were out in full force trying to determine exactly how the fire started. This they knew: The fire was caused by a person, according to Ruth Wenstrom, a Forest Service spokeswoman.

"Earlier this summer, we had a very low rate of fire starts," said arson investigator Ron Huxman when the Bridge Fire was over. "But in recent weeks, we saw a very definite increase in the number of man-caused fires, particularly those caused by arson."

There were several confirmed arson fire starts in the Cajon Pass area in the weeks leading up to the Bridge Fire. The series of suspicious fires happened about ten miles west of where the Bridge Fire erupted, and the fact that the fires occurred in the Cajon Pass did not pass unnoticed by investigators. In fact, the Cajon Pass contains a major rail artery and handles more rail freight tonnage than any other mountain pass in the world. It is also one of the busiest for road traffic, being the only major route to Las Vegas from the Los Angeles area and the surrounding counties. A fire roaring through the pass in hot, scorching conditions—accompanied by unchecked winds which are common that time of year—would spell disaster for any motorists transiting the area, as there are few feeder roads or exits leading into safer areas.

Also, near Cajon Pass, three brush fires suspiciously broke out at the same time along Highway 138, near Interstate 15, on the afternoon of August 28. The largest fire burned twenty acres, and investigators suspected arson in all three. The acreage consumed in the Bridge Fire totaled 1,352 acres.

California Department of Transportation (Caltrans) road crews spent Tuesday, September 9 replacing nearly four thousand feet of guardrail damaged in the fire, and Southern California Edison personnel had to replace many burned power poles. Crews also cleaned the highway.

The fire had burned through thick brush, and most of the area was blackened and denuded all the way to the ground, setting the stage for possible severe flooding when and if rains returned. As it turned out, runoff from the barren Bridge Fire area was responsible for major rock slides and mud slides during the winter of 2003–2004 that several times damaged and closed Highway 330, which is a major artery into the mountains.

At the height of the fire, as flames moved toward the communities of Running Springs and the Lake Arrowhead Resort area, about fifteen hundred homes and three thousand outbuildings were threatened.

In spite of the path burned by the fire as it came up the hill, the danger of a catastrophic fire in the San Bernardino National Forest remained extremely high due to the previous four years of drought and the bark beetle infestation that had killed thousands of trees. It was the fast response of firefighters that was credited with slowing the blaze the previous Friday and Saturday.

Karen Terrill, a fire information officer for the CDF, said, "The last time this area burned was 1956, so this is a lot of very dead, very dry material that burned, in addition to the bug kill area," she said.

Another blaze, called the Lilac Fire, consumed 170 acres near the Pala Indian Reservation in rural northern San Diego County during the weekend of the Bridge Fire.

The Lilac Fire moved in the direction of the reservation,

an area that had no history of burning. "So what's burning there now is very dead and very dry," Terrill said.

In an interesting twist of fate, the burn path of the Lilac Fire would later help firefighters battling the disastrous Cedar Fire in October. The Lilac Fire burn path would divert flames away from a group of homes bordering the reservation and would give firefighters time they would need to establish a strong line before a wind change that would threaten the tourist town of Julian in the Laguna Mountains northeast of San Diego. "It was just like we had set a backfire, right where we would have set a backfire," one firefighter from Camp Pendleton Marine Corps Base would later recall.

BRIDGE FIRE STATISTICS

The Bridge Fire started at 3:04 P.M. on September 5, 2003, in the vicinity of the City Creek Fire Station on Highway 330, about ten miles southeast of Lake Arrowhead in the San Bernardino National Forest. The Bridge Fire was turned back over to the San Bernardino National Forest and a management team on the morning of Thursday, September 11, 2003.

Fire rehabilitation, a process of helping an area heal more quickly after a fire and prevent further damage, began at that time with the cleanup of areas impacted by the fire operations. A Burned Area Emergency Rehabilitation Team continued working with hand crews the week following the fire. They studied the burned area and created prescriptions to mitigate damage and prevent future damage from winter rain. Months later in this area, helicopters dumped many tons of hay on the hillsides to help hold the soil and prevent dangerous runoff.

Crews were released to their home units or to fires that were burning in other parts of the country. The Incident Command Post at the San Bernardino County International Airport was disbanded as the incident was declared over.

Initial Attack Incident Commander Mick McCormick reported the fire as five acres upon arrival Friday afternoon.

By 8:30 in the evening the fire, which was called "erratic" by Forest Service Information Officer Tricia Abbas, was reported to have consumed about fifteen hundred acres. Humidity of around 50 percent was causing the fire to "lay down," or calm, slowing the rate of spread. Sixty-five engines and twenty-four hand crews made up a total manpower contingent of more than five hundred firefighters, but more strike teams were called in. There were six strike teams on the mountain.

Because the fire was erratic, it left many unburned islands of fuel. Crews did some burning-out operations when conditions were cooler, with higher humidity, to reduce the unburned areas. Fifteen hundred homes were in the path of the fire threat, with the leading edge of the fire only about one mile from the mountain community of Fredalba. Nearby Smiley Park and Running Springs were also threatened.

USFS Mountaintop District Ranger Allison Stewart stressed that due to the intense public information program about the high fire danger, and fire safety in these areas, "The public is scared to death," and she asked that all firefighters be sensitive to this when dealing with local residents. Reports confirm the firefighters were extremely aware of the public's fear and kept that fact in mind whenever they interacted with residents.

Between eight hundred and one thousand residents were evacuated to Rim of the World High School. However, the Red Cross stated that the high school could not handle that many evacuees, and another location was sought that could accommodate more people.

The evacuations were mandatory for Fredalba and Smiley Park and voluntary for Running Springs. The San Bernardino County sheriff executed another evacuation sweep on Saturday morning. The Mountain Rim Fire Safe Council was involved with communicating conditions and instructions to the residents.

Only one outbuilding was lost, and the command team was determined to protect all the homes in the area, no matter how small.

While the fire was fought from the ground by engines, hoses, and hand crews, twenty aircraft waged an air assault. Fixed-wing air tankers dropped Phos-Chek fire retardant, and helicopters dropping water taken from Lake Arrowhead were utilized until dark.

The incident was declared officially over on Monday, September 15, 2003 at 3:55 P.M., at what would be a total incident time of 240 hours and 51 minutes.

The fire burned about three square miles of thick chaparral, which is home to many native birds and mammals. Much of the area was on steep watershed, creating the risk of downstream urban flooding and possible loss of riparian habitat farther down City Creek, which flows through that area and down to the Santa Ana River in the valley below.

FIVE

The Honorable Ann Veneman

"Astonishing to see . . . ,"
*Agriculture Secretary Ann Veneman
after flyover*

A month before the Bridge Fire, a significant event had taken place in the San Bernardino Mountains. On Friday, August 8, 2003, a pine green helicopter from the U.S. Forest Service ferried in representatives from federal, state, and local government to the Lake Arrowhead landing pad at the Mountains Community Hospital for an up-close walking tour of a forest in need.

Shortly after landing, Secretary of Agriculture Ann Veneman, along with Congressman Jerry Lewis, State Senator James Brulte, and San Bernardino County Third District Supervisor

Dennis Hansberger were greeted by officials of the county fire department, U.S. Forest Service, and CDF. They were also met by a large contingent of concerned citizens who were looking to these elected officials to do something to prevent a fire of titanic proportions that many feared was imminent.

Ann M. Veneman—ninth in the line of succession to the presidency—had been unanimously confirmed by the U.S. Senate and sworn in as the twenty-seventh secretary of the U.S. Department of Agriculture on January 20, 2001. During her Senate confirmation hearing, she repeated President George W. Bush's belief that "the spirit of the American farmer is emblematic of the spirit of America, signifying the values of hard work, faith, and entrepreneurship." Secretary Veneman pledged to work to foster an atmosphere of teamwork, innovation, mutual respect, and common sense within her department. Her ideas were certainly being put to the test on this day.

Veneman served as U.S. Department of Agriculture deputy secretary from 1991 to 1993. From 1989 to 1991 Veneman had served as deputy undersecretary of agriculture for international affairs and commodity programs.

An attorney who was raised on a peach farm in Modesto, California, Veneman earned her bachelor's degree in political science from the University of California at Davis, a master's degree in public policy from the University of California at Berkeley, and a Juris Doctorate degree from the University of California, Hastings College of the Law.

Upon disembarking from the helicopter, Veneman and the other dignitaries received a short briefing by Bob Sommer of the Forest Service concerning the various types of decay and a description of the areas hardest hit by drought and the bark beetle. Sommer pointed to a hillside covered completely with dead brown pine trees.

"To give you an idea of the dynamics of the situation," Sommer said, "six months ago if you looked this way at this hillside it would have been completely green. The north-facing slopes took a little bit longer to die because the sun doesn't shine on them so much, but now that the bark beetles are well established they can kill even the healthiest trees— like the ones we have in the shaded areas."

The secretary and the officials were then driven by motor-cade to a neighborhood on John Muir Road, where they watched a tree being felled with the help of a large crane. The tree was the responsibility of Southern California Edison, whose representative explained the protocols for de-energizing the power lines when such close-haul tree removal operations happen. Veneman asked how long the power would be off and whether the residents were given a window of time to plan for when power would return. The Edison representative, David Simmons, informed the secretary that residents are always warned of impending power shutoff when lines were to be de-energized or lowered for tree removal.

He also said the cost can run as high as ten thousand dol-lars a day for Edison crews to perform line de-energizing and obstacle removal. When asked how many trees can be felled during such downtimes, Simmons replied, "Between fifteen and twenty in this kind of situation."

"Look, look!" Veneman declared, as the tree was lowered gently to the ground by the crane after being cut. After they had a chance to inspect the downed tree, Veneman and Lewis took the opportunity to speak with some area residents. "I'm shocked at the situation up here!" Veneman exclaimed. The secretary seemed in awe of the work the tree cutters performed and complimented one worker on a job well done.

The motorcade then proceeded to a location along Innsbruck Drive on the northwest shore of Lake Arrowhead where a grove of approximately five hundred dead and brown trees was encircled by homes.

Veneman and the other officials listened intently as Tom Bonnicksen, Ph.D., a Texas A&M professor of forest sciences and widely recognized forestry expert and lecturer, explained ways a healthy forest can be planted after the dead trees are removed.

Bonnicksen pointed to a nearby stand of dead trees. "We need to prevent this from happening in the future," he said. "There is virtually nothing left. The forest is gone. You'll see a hillside with a few oaks, and to recover from that—without doing anything to replant it—would take, in my estimation, a century or maybe even two centuries."

Veneman appeared genuinely dismayed by Bonnicksen's report, asking what, if anything, might be done at this point. "If we do something, we can have a forest back here—it may be a young forest—by removing the dead trees, and thinning the pockets, then use native species to replant," Bonnicksen insisted.

Veneman listened to Bonnicksen. "We can accomplish this," he said. "The community says they want to do it, we have the know-how, but when we bring this forest back, it can't be one like a hundred years ago; it has to be one that is sustainable."

At the June 9, 2003, public meeting sponsored by the Arrowhead Communities Task force and the fire safe council, Bonnicksen had presented the potential of great achievements with the work in the San Bernardino Mountains. "What happens here sets a precedent for what will happen elsewhere," he said. "If we can deal with this problem here effectively, thoughtfully, and with a long-term perspective, I think we can do so elsewhere as well."

At a later interview conducted at the *Mountain News,* Bonnicksen discussed his opinion concerning the way the government has allocated forest funds.

"When you sprinkle four hundred million dollars around the West, you get nothing," he said. "In my view, the money has to be concentrated in those places that need the help now, and a long-term plan has to be carried out in that area to demonstrate that this will work to solve the problem. And in my judgment, we're not solving the problem."

After listening to Bonnicksen and viewing the decimated forest stands along Innsbruck Drive, Veneman returned to the helipad and bid farewell to her hosts, thanking everyone for the enlightening tour.

The secretary boarded the waiting helicopter and was whisked away to a press conference at the USFS Air Tanker Base in San Bernardino. It was there, fire officials, firefighters, and Forest Service officials hoped, the poised and charming Veneman would make some sort of announcement indicating the government saw the scope of the problem and was prepared to take action and do something about it.

"I am gratified that Secretary Veneman, a native Californian who understands the dangers of forest fires, agreed to see firsthand the terrible devastation that has come to our Southern California forests," Congressman Lewis said at the post-tour meeting. "It is clear that the secretary and the Forest Service are making a top priority of the need to reduce the fire danger caused by a million dead trees in and around our mountain communities."

Then, clearly shocked at the fire danger posed by dead trees in and around private and Forest Service lands, Veneman announced at the press briefing that the Forest Service

LONGER FIRE SEASONS MEAN MORE FIRES

Warming and climate change are the major reasons for a quadrupling of the number of large wildfires in the western United States, rather than land-use and forest-management practices. That is the conclusion of a study published in *Science* in August 2006.

Researchers from the Scripps Institution of Oceanography in La Jolla, California, including climate authority Anthony L. Westerling, studied thirty-four years of western wildfire and climate history since 1970. They found that the number of large wildfires (more than one hundred thousand acres) began to substantially increase in the 1980s in concert with a lengthening of the fire season, caused by increasingly early snowmelt and other factors associated with climate warming.

They concluded that earlier snowmelts can herald

had secured an additional $5 million to help fight the ravages of the bark beetle infestation in the San Bernardino National Forest.

"With these additional funds, we have more than tripled the San Bernardino National Forest's original budget for this essential work," she said. Veneman said the Forest Service would take a two-pronged approach to solving the problems.

The first approach would involve supporting President Bush's Healthy Forest Initiative.

longer dry seasons, which create a longer period in which large fires have opportunities to start. Earlier snowmelt can also mean greater drying of vegetation and soil, which feeds fires and makes them much more difficult to contain. The researchers noted that climate model projections show springs and summers throughout this region will continue to warm in future decades.

"These trends will reinforce the tendency toward early spring snowmelts and longer fire seasons," the team wrote. "This will accentuate conditions favorable to the occurrence of large wildfires, amplifying the vulnerability the region has experienced since the mid-1980s."

Source: A. L. Westerling, H. G. Hidalgo, D. R. Cayan, and T. W. Swetnam, "Warming and Earlier Spring Increase in Western US Forest Wildfire Activity," *Science* (August 18, 2006), vol. 313, 940–943.

"The president is committed to the issue of restoring forest health and preventing the catastrophic fires we have experienced the past few years," the secretary said.

The second approach would involve continuing efforts to remove fuels from the forest. These fuels have been blamed for helping to sweep fires along once they've taken hold.

Responding to a question from CBS News reporter David Garcia, Veneman said she considered the situation in the local mountains "very serious" because the affected areas were full of fuels and noted the mountain communities were basically surrounded by the dead trees on Forest Service lands.

When asked by one reporter why a national emergency declaration hadn't been made, Veneman explained that a declaration of national emergency is issued by the Federal Emergency Management Agency. Lewis and Veneman discussed the fact that FEMA is not set up to help prevent a disaster ahead of time. Instead, FEMA is set up to respond to emergencies once they have occurred.

"I understand the severity of the fire threat. Declaring a national emergency is under consideration by the president," Veneman added, and said that after seeing all the damage, she was going to discuss the seriousness of the situation with President Bush when she met with him later that week.

THE ENEMIES: DROUGHT AND A TINY BEETLE

Many people are in denial that the Earth is rapidly warming and that humans are partly to blame. I am not one of them.

In September 2004, the National Geographic Society threw caution to the wind and dedicated seventy-four pages of that month's issue of *National Geographic* to the truth about global warming, potential canceled memberships to the Society be damned. In that issue, editor in chief Bill Allen wrote:

> I'm asking you—even those of you who don't believe the Earth is getting warmer and that human behavior is a contributing factor—to [recognize] this isn't science fiction or a Hollywood movie. We're not going to show you waves swamping the Statue of Liberty. But we are going to take you all over the world to show you the hard truth as scientists see it. I can live with some canceled memberships. I'd have a harder time looking at myself in a mirror if I didn't bring you the biggest story in geography today.

And this story has grave implications for the threat of devastating wildfire. Computer models show that carbon dioxide–induced warming could increase the chance of devastating fires in northern coniferous forests by more than half.

The danger is growing in the Southern Hemisphere as well. Unprecedented wildfires broke out in the Brazilian rain forest last year, threatening a vast area that many scientists believe is vital to the health of the entire Earth. Increasing peat fires over wide areas of Indonesia could also be contributing to global warming by pumping great amounts of carbon dioxide into the atmosphere, while shrinking rain forests provide fewer trees to soak up the increased greenhouse gas.

That is just the effects of the warming, fires, and drought conditions. It doesn't take into account other factors that are exacerbated by heat and drought, such as bark beetle infestation. What happens to fire potential if huge areas of the forest are dead?

The Southern California bark beetle infestation received widespread media coverage, as agencies and people tried to take down millions of dead trees before they could be consumed by wildfires. However, similar disasters were happening in other places as well, and in some on a greater scale.

On the Kenai Peninsula of Alaska since the early 1990s, the spruce beetle has laid waste to an area the size of Connecticut. And the area of destruction there keeps growing.

Bark beetles are a part of the forest ecosystems throughout the West and always have been. There are six thousand species of them worldwide. Every year, they kill trees as they bore their trails under the bark of vulnerable trees. It is only when an outbreak occurs that people take notice.

A bark beetle is tiny, only about the size of a grain of rice, but millions of them concentrated in one area can do great damage as they eat under the bark of trees made vulnerable. Drought and heat speed up their activity and allow them to survive through winter. Precipitation and cold normally cause the majority of a population to die back. During warm years, there have been outbreaks, but generally, North American winters have been too cold and wet to support lasting outbreak populations of bark beetles.

That began to change in the 1990s, as the world began heating up and drought took hold in many places in the West. The spot infestations of bark beetles that occur naturally began exploding into outbreak populations, destroying vast forest areas in California, Utah, Alaska, and other states.

The bark beetle kills trees by boring into the tree's vascular cambium layer, often making a complete ring under the bark of a tree, robbing it of moisture from the roots below. The tree characteristically "dies from the top down," first turning brown at the top and then rapidly dying. Once a tree is dead, the bark beetles move on to a living tree and begin their work again. Healthy trees with ample water and loaded with sap are less vulnerable to the beetle, partly because they are able to produce resin that pushes the beetles out of the holes they bore under the bark and also often traps and suffocates them.

The trees in the Southern California mountains were not able to repel the onslaught as bark beetles raged unchecked for years. During drought many trees can survive infestation if the forest itself is "healthy," with widely spaced trees. In Southern California, however, the forest was seriously overcrowded and trees had to compete for precious underground water. The closeness of the trees widened the disaster as bark beetles flew easily from one tree to the next.

Drought itself can effectively kill large areas of trees, but when it is combined with infestations such as these, vast tinderboxes can be quickly created. This became evident as the October winds began blowing in Southern California in 2003.

SIX

The Calm Before the Storm

My job as the editor of two weekly newspapers on the mountain was all consuming. Each week, just when we'd put one newspaper to bed, my editorial staff and I would face the prospect of staring at the blank page, and then we would start all over again.

Since assuming my post in December of 2002, I had not taken any personal time. It was just too much of a hassle, and besides, every so often my family and I would take off on four-day weekends right after the papers hit the streets on Thursday mornings.

One such four-day holiday, however, would be scheduled for an unhappy reason. My sister-in-law had finally succumbed to injuries she sustained while horseback riding on a holiday in Montana the previous July. She had been in a coma since then, and she died on Thursday, October 16, 2003.

We left right away for Fort Collins, Colorado, in a rented Ford Windstar minivan.

With me on the trip east were my mother-in-law, Laverne, my wife, Karen, and our four-year-old daughter Gracie. Our older daughter, sixteen-year-old Rachael, and our sons, Richie and Jake, would remain in the San Bernardino Mountains with Karen's dad, Harold, while we were off to Colorado. Since Harold was busy with his nursery business and Rachael had just started a new job as a server at one of the most popular diners on the mountain (and the boys had previous plans), it would be just us four out on the road.

After enjoying a buffet dinner at one of the glittering casinos along the revamped Las Vegas strip, we stayed the night in St. George, Utah, and rolled into Fort Collins late the next day.

The funeral, while sad, was unlike any I had ever attended. It was a celebration of a life too short-lived. After spending another night with my brother-in-law, Ken, we left Friday morning, October 24, for the long trip back home.

Friday night was spent in Las Vegas, New Mexico, a town far different from the Las Vegas we had visited on our way to Colorado. The little city was a ghost town, with the entire population sitting in the high school football stadium. It was homecoming night, and merchants had closed their businesses early to enable the owners and employees to cheer on the home team. Signs were peppered everywhere across town: "Gone to Homecoming, see you tomorrow!"

Despite all the festivities, we managed to find a Dairy Queen that was open and took our dinner back to the motel.

The next morning, Saturday, October 25, 2003, our departure was delayed because of a flat tire on the Windstar. A couple of hours later we were on the road again, and then came the call.

"Mom, it's Rachael," she said with a tremor in her voice. Not quite fear yet, but definite concern. "Mike Neufeld called. He said there's a fire in Waterman Canyon and that houses are burning in Crestline and they're afraid it might burn Arrowhead, too." Karen calmed her as I reached for my own cell phone and got Mike on the line.

Neufeld was my chief correspondent at the *Mountain News*. A reliable old dog of a reporter, Mike possessed a news background that included long-term stints at the Associated Press, and television news work in San Jose for Fox. He had semiretired in Lake Arrowhead and found his duties as a reporter on the mountain filled the bill perfectly for him.

"Mike, what's up?" I asked, as we drove nearly 80 miles per hour across the New Mexico plains.

"It's bad," he said. "They're saying about ten or fifteen houses have burned so far and it's just getting worse."

According to Mike, the winds were fierce, which fueled the fire even more, and the worst part was the flames were headed up the south face of the mountain. All of the disaster models predicted the flames would come from the north, fanned by the infamous Santa Ana winds.

With the winds blowing the flames both down and up the mountain, the "perfect storm" conditions firefighters so feared had materialized. There was no stopping the conflagration.

When I ended the call with Mike, I quickly dialed Peter Brierty's personal cell phone. I knew he'd be in the thick of things, and I was right. He described the valiant effort the firefighters made in Crestline, but his tone of voice gave me concern. This was also the call when I first learned the name the incident commander had selected for the fire. The fire was officially entered into the annals as the Old Waterman Canyon Fire. For brevity's sake, by midafternoon everyone would be referring to it as the Old Fire. The name stuck.

Throughout the day as we passed through Santa Fe, Albuquerque, and Gallup Brierty called with updates. Each call brought more bad news.

I had also managed to reach our Crestline area correspondent, Lee Reeder, who had heard the numerous sirens and was on the way to the Rim of the World area along Highway 18—featuring a number of turnouts and scenic viewpoints—minutes after the fire broke out in lower Waterman Canyon. Lee had some of the first photographs of the fire as it raced up the mountain, and the photos were terrifying.

Huge plumes of smoke towered over the canyon, and the flames were clearly visible as they jumped Highway 18 at will. Another fire was sparked, which firefighters dubbed the Playground Fire, named for nearby Playground Road, late that afternoon. By evening the two fires had merged, with one finger headed toward the mountain community of Cedarpines Park and the other bent on consuming as much of Crestline as possible.

Following a powerful onshore wind shift, a third finger

began to reach for the town of Rimforest. In the late afternoon, when we stopped at a Wal-Mart in Arizona to replenish our batteries and cell phone cards, I caught a glimpse of the fires in a photo displayed on the front page of *The Arizona Republic*. The Grand Prix Fire had been burning unchecked since Tuesday, and true to the old saying, the picture said a thousand words.

It would really hit me a few hours later when we pulled into Flagstaff, Arizona, for the night.

As we checked into the Travelodge, the night clerk at the desk had tuned his television to CNN. There it was, in living color. The cruel and deadly flames were enjoying the fifteen minutes of fame afforded by Larry King. As we checked into the motel, we were transfixed by the sight of our town, our mountain, going up in flames. The Big One had arrived.

SEVEN

Personal Stories

On the day before we got the news about the Old Fire, the Grand Prix Fire, burning about fifteen miles to the west of Waterman Canyon, was forcing thousands of people from their homes. Early on the morning of Friday, October 24, Janine McCord of Hunter's Ridge in Fontana, had about twenty minutes to evacuate four children, two dogs, and their most important belongings. That included her computer, birth certificates, medicines, a breathing nebulizer for her daughter, and a supply of dog food.

McCord, standing outside the Jessie Turner Community Center in Fontana, was most grateful her children were safe, knowing all the rest was replaceable. That Friday the Grand Prix Fire was displacing hundreds of residents who sought shelter and information at evacuation centers in Fontana and Rancho Cucamonga.

The American Red Cross organized both of those shelters, which were bolstered by an outpouring of support from the communities. With helicopters roaring overhead and flames visible from the northeast corner of Rancho Cucamonga High School, people gathered throughout the day at that evacuation site to hear the latest news about the fire.

By 5:00 P.M., 250 people from the northern parts of Rancho Cucamonga and Fontana checked in at the high school gym, though many were planning to stay at a hotel or with family if they couldn't return home that night. As soon as evacuees poured in, donations and volunteers arrived.

By lunchtime local grocery stores, Calvary Chapel in Rancho Cucamonga, Target, Subway, Papa John's Pizza, and Coca-Cola had donated more than enough food. "Anything left over will likely go to the American Red Cross," reported Steve Reinelt, a Red Cross disaster manager.

Many people brought their dogs and cats to the evacuation center, requiring San Bernardino County Animal Control people to set up an animal shelter on a nearby baseball field.

Sergio and Isabell Villanueva, with their daughter, Christina, evacuated their Hunter's Ridge home early Friday. When they awakened that morning it was oddly dark and hot from the flames and smoke. They knew it was time to go.

Now the family sat with their dog and cat outside the gym and watched the smoke cloud grow. A reporter from the San Bernardino *Sun* cornered Isabell as she returned from the food table and asked her why she chose to remain at the evacuation center when her home was really blocks away from the crux of the fire.

"Even if we wanted to stay, I don't think it would be safe," Isabell replied.

The reporter dutifully logged her words, but he was really looking for someone who could give him the kind of breathless quote his editor would expect when he filed the story later in the day.

While volunteers from the American Red Cross and the city's Community Services Department organized the instant shelter, it was business-as-usual-as-possible at the school, where half the students were either absent or helping with the evacuation center.

"It's going as smooth as it can," Rancho Cucamonga High School principal Matt Holton had assured the day before. But the women in the attendance office called out dozens of names over the loudspeakers to pull students out of class. There were too many "parental requests to leave" to send individual notes to classrooms. The parents wanted their kids.

Susan Winckler, a math and music teacher, was able to run some classes normally, but a few were half-full, so she had to improvise.

Tom and Cindy Dupart, with their son, Tyler, learned the night before they would likely be evacuated. They slept little, fearing the fire. They packed up belongings: pictures mostly, some clothes, toys and juice for Tyler. For the most part, they just sat tight, playing the waiting game.

By about 6:00 P.M. on Friday, October 24, the Red Cross emergency shelter at the Jessie Turner Community Center had signed in 109 adults and fifty-four children in the last twenty-four hours, recalled Diane Dyan, a Red Cross volunteer and site manager. Many more were expected to spend the night on Friday. About thirty people, mostly Lytle Creek evacuees, spent Thursday night on cots. Red Cross volunteers provided "comfort kits," with teddy bears for the children. A message board was created for evacuees.

One woman brought boxes of homemade sandwiches, and several more businesses donated pizzas, milk, cookies, and bottled water.

A striking blue and gold macaw sat on Doug Chapdelaine's shoulder in front of the Fontana shelter on Friday. *It's terrible*, Chapdelaine thought. He told anyone within earshot, "If it's an arson, they ought to put him away for life. It's a sickness they have; they need to be put away."

Neighbors helped Doug's wife, Shirley, evacuate their dog and two birds after Doug left for work. They later met up at a restaurant and went to the shelter together. Shirley Chapdelaine said she had just enough time to put her animals in the car, get medicines, and grab her wedding ring and important papers. Somehow, she said, she even managed to remember her Rolodex.

On Monday, Mary Anne Craft of Claremont took a walk to survey the fire damage around her High Point Drive neighborhood. To her surprise, three units had been destroyed two streets away. *I didn't realize that there was any damage up there. I was shocked*, Craft remembered thinking.

While she was walking by the burned-out units, a photographer from the *Daily Bulletin* snapped an action shot of Craft. She gave the photojournalist her permission to use the photo for publication. What happened next she couldn't have possibly imagined.

Her brother, who lives in Rochester, New York, was on a business trip in San Jose and tried to call her after hearing about the fires in California, but he couldn't get through because phone service was out in Claremont. Then, purely by coincidence, he purchased a copy of the San Jose *Mercury News*, and was amazed to see a photo of Craft walking in front of the burned-out dwellings. He nearly fainted. He called his wife in New York to try phoning or e-mailing Craft, which she did. Fortunately, phone service had been restored and all of Craft's family members were soon assured of the well-being of both her and her house.

Meanwhile, that same day, about 750 goats descended on the still fire-free Claremont hillsides to eat the grass and brush, a measure to help stop more wildfire outbreaks. The Los Angeles County Fire Department hired the unusual work crew. The goats can clear fifty acres of brush in approximately one month. Goats are less invasive than chain saws or other heavy machinery, and goat use is one of the many California Firewise Management Practices endorsed by the County of Los Angeles Fire Department.

As fires continued to rage in the Inland Empire, the son of a former San Bernardino County supervisor remained in critical condition after he was burned when the Grand Prix Fire engulfed his car as he was helping his parents evacuate. Mickey McElwain was overcome by flames as the fire bore down on his parents' house.

After packing the cars, McElwain and his mother, Nita McElwain, were trying to get out of the driveway. Mickey

McElwain went first, then a forty-foot pine tree fell, blocking Nita McElwain's exit from the driveway. It took her about two minutes to maneuver her way out, and by the time she was in the road she turned to look for her son and noticed a burned-out car. Cal McElwain began to call for his son, realizing that it was Mickey's car that was destroyed. Out of the smoke, Mickey McElwain stumbled into his mother's car and told her that he was burned.

More flames rose in front of them. Now there were flames behind and in front of them. They put the car into low gear and floored it through the wall of flames.

Cal and Nita McElwain would later say they were not given adequate notification to evacuate. Their home was destroyed in the fire.

Susana Acosta was walking to her car at Emerson Middle School in Pomona when she came upon a picture of a man in uniform in front of a barracks. The photo was lying faceup in the street, and when she picked it up Acosta read an inscription on the back of the photo: "Mic McElwain, Japan."

The teacher's aide is a student at Mount San Antonio College. She showed the photo to her teacher Shane Palter during a discussion they were having regarding a grade. This was on Wednesday evening, October 29.

"McElwain is an unusual last name, so we just got out the phone book and the first person we called was the daughter of the man in the photo," Palter said. Cal McElwain, whose nickname is Mic, served in Japan after World War II. The photo was taken in 1946, when he was seventeen years old. All the rest of the pictures in the boxes were destroyed, but the one surviving picture was found by Acosta almost fifteen miles away. It had blown there on winds generated by the fire. "I think finding this picture was a blessing," Acosta said. "I really hope to be able to meet the family and talk to them."

CEDAR GLEN, SAN BERNARDINO COUNTY, OCTOBER 29, 2003

Hugh Campbell of Cedar Glen recalls all he could do was dive for cover when he saw the firestorm racing toward him. "I just hoped for the best," he remembers.

The inferno came fast. Embers rained down as massive amounts of fuel encircling him were consumed by the hellish blaze roaring toward him. He felt the heat singe his hair, and he began to choke on the smoke that at first only resembled haze. Scrambling for a shelter he'd created out of cinder blocks and sheets of concrete, Campbell ducked inside without a second to spare. The powerful flames rolled over him in a massive fury, and he thought for a moment, *This is it. I'm going to die.*

Campbell emerged to find what appeared to him to be a charred battlefield. The fire would continue burning everything in Campbell's sight for at least another twelve hours. His house and homes that belonged to two of his neighbors were spared.

"It was a hell of a fire," Campbell would recall. "A hell of a fire."

The three homes were among the few in Campbell's neighborhood to survive the inferno that laid waste to three hundred Cedar Glen homes. The next day firefighters would find what they would describe as "an alien landscape of charred trees, smoldering rubble." They would also discover Campbell as he walked the area around his home in stunned disbelief.

"It takes a combination of stupidity and luck to do what he did," one firefighter observed. "I'm a fireman and I wouldn't have done it."

Campbell, an aerospace engineer and Navy veteran, knew the fire would come, but he was set on defending the home he had bought in 1989. He stayed behind when the evacuation order came earlier that week, but he made the mistake of heading down the mountain Monday for gas. The California Highway Patrol wouldn't let him return home. "We can't let you through," they told him. "It isn't safe up there."

Campbell would not be deterred.

"I knew they wouldn't save this place," he said, nodding toward the tidy brown cabin he called home each weekend. The rest of the week was spent in Anaheim, where he worked at a defense firm. "I knew I'd have to."

He sneaked home Tuesday along unblocked back roads. He cleared brush and felled trees. He dug a firebreak and shut off the gas main. As a final precaution for "the worst-case scenario," he built a small shelter to protect himself from the flames should they come for his house.

"And then I listened to a Mendelssohn violin concerto and waited for the fire," he said, wiping ash from a dirty green jacket.

The fire came just after noon Wednesday, through the canyon just beyond Campbell's house. Campbell saw it through his binoculars. *Oh good,* he thought, *it's moving away from me.*

Then the wind changed direction. The fire was moving toward Cedar Glen.

Campbell grabbed his hose and doused his house before turning the water on his neighbors' homes. Then he started digging another firebreak. He never finished it. The fire moved surprisingly fast.

He shakes his head at the memory. His eyes were red from smoke and ash, which filled the air and coated everything in sight. He says he was never scared, but his face reflected the emotions of someone who had faced death.

"It's just a matter of believing in yourself and having a plan," Campbell said. "Once you commit yourself to something, the fear just disappears. As a result, I'm the only house still standing." Along with his two neighbors'.

"Yeah," he said with a laugh. "And those two guys are going to owe me a nice dinner when they get back."

HARRY BRADLEY, PUBLISHER OF THE *MOUNTAIN NEWS*

The following extract is a personal account of the newspaper staff efforts during the Old Fire written by *Mountain News* publisher Harry Bradley. As the managing editor of the *News*, I did my best to stay away from all the hyperbole and the many nonfactual reports that were emerging from this constantly changing story. This could be difficult at times, especially knowing that our own homes were in danger.

It became an ongoing challenge to remember that whenever the true facts of any situation are elusive or missing, as human beings we seem to have a natural propensity to fill in the blanks with any piece of information at hand no matter the source. That is why receiving Harry's calm, factual report meant so much to me personally during those terrible first days as the fires raged unchecked. Here is Harry's report:

On Saturday [October 25, 2003], I was covering the Arrowhead Lake Association meeting at the Burnt Mill Beach Club. The new board of directors was being installed.

Director Chris Klein approached and whispered, "There's a fire in Waterman Canyon."

County and state fire departments and Forest Service officials were gathered at the Crestline turnoff, looking down at the first stage of what we would come to know as "the Old Fire." Plans were immediately made to combat the fire and begin evacuation of Waterman Canyon and possibly Crestline.

I drove down into the burning canyon and witnessed the cooperative efforts of CDF and San Bernardino County firefighters as they saved several homes along Old Waterman Canyon Road. The wind was 33 mph at the Crestline turnoff but it blew as high as 50 to 60 mph down in the canyon. In fact, the wind blew my hard hat off, and it flew into the fire. Rocks fell onto State Highway 18 as the fire burned vegetation.

Cellular phone calls and messages began flying among our news team. Our editor, Davey Porter, was driving home from a family funeral in Colorado but was in contact with San Bernardino County Fire Marshal Peter Brierty, who kept our reporters informed of the situation as it progressed, while our associate *Courier* editor, Lee Reeder, and reporters Mike Neufeld and Joan Moseley went into action.

I telephoned my wife, *Mountain News* Advertising Representative Angela Yap, and told her to prepare for evacuation to our "little house" in the desert. I felt that evacuation and power outage was imminent. We left the mountain just before sundown, and formal evacuation began in the wee hours of Sunday morning.

Sunrise Monday morning I picked up Davey Porter from an Apple Valley motel and we made the drive up the back side of the mountain on the dirt road portion of Highway 173. Our press credentials allowed us entry, and we arrived on the mountain to find an eerie, almost surrealistic silence.

Myles White, one of our graphic artists, had slept through the evacuation the night before and had ridden his bicycle to the office. After we started up the generator to run our newsroom computers, sheriff's deputies arrived and advised us to leave the mountain.

We packed up the computers, servers, hardware and software into the delivery truck. Davey and Myles met our graphics production manager, Pete Noriega, at our company's printing plant located at our sister newspaper, the *Hi-Desert Star* in Yucca Valley.

By sundown the news team was updating our newspaper's Web site with the latest news, and reporters continued filing their stories via e-mail and cell phones.

Advertising consultants Angela, Lisa Langdon and Wendy Blackburn drove out and worked at the Yucca Valley office. The *Mountain News*, the *Crestline Courier-News and the Big Bear Grizzly* were able to publish due to the incredible cooperative efforts of Brehm Communications' Southern California cluster of newspapers and the printing plant.

The *Big Bear Grizzly* staff was evacuated Tuesday afternoon and the *Hi-Desert Star* also provided them emergency office space. Judi Bowers, editor of the *Grizzly,* filed reports from the command center located at the Big Bear City Fire Department.

Our news team rented quarters at the Sky Valley RV Park, eight miles east of Desert Hot Springs. We were grateful to our friend Susan, an Arrowhead resident who works at the RV park, for her help in arranging the rentals.

On Tuesday, October 28, I again made the trek up the Highway 173 dirt road. Wearing the protective yellow gear, which included a hard hat identifying me as "PRESS," enabled me to move throughout the town with my camera.

Many of the fire captains recognized me from the Town Hall Meetings, and I was usually greeted with "Hi, Harry, what the hell are you doing here?"

Captain Wayne from Lake Arrowhead station 91 looked like he hadn't slept since I saw him on Saturday morning. He was holding a meeting on the hood of his SUV with other fire team leaders. I unloaded all the cans of iced tea and water out of my cooler and placed them on the hood of the truck and wished them luck.

I talked with Butch Bauman, owner of Rim Forest Lumber. He said the firefighters saved his lumberyard Monday night by spraying gel on everything. Southern California Edison's station was on the point of devastation in Rimforest.

It was spared and the other commercial buildings behind it were OK. There was a hole burned in the shake roof of the rental store but it was extinguished before the fire spread.

Numerous buildings west of Edison had burned to the ground, including Charter Cable, NeuArt Studio, the cabinetmaker's shop, and several homes on the hill above Highway 18. The gas company was on scene to cap the exposed lines that were still on fire.

The police were stopping everybody and checking them out. Nobody was to be in the mountains; however, two of my neighbors—both retired firefighters—were holed-up with generators, satellite TV and fire hoses ready to hook up to the fire hydrant and hopefully save their homes.

The firestorm was being fought along Highway 18 across from Rim of the World High School, all the way to Santa's Village.*

Skyforest and Willow Woods were saved; however, flames demolished two cabins on the downslope of Highway 18 towards Santa's Village.

The fire crossed the highway and burned Wiley Woods Conference Center but they were able to extinguish the giant dead pines in the meadow behind Santa's Village before it spread towards Lake Arrowhead.

Across the highway from Santa's Village, stacks of bark beetle logs were ablaze. Airplanes skimmed the treetops. The big bombers were guided in by smaller planes in an or-

---

* Santa's Village is a now-closed wintertime amusement park located in Skyforest, near Lake Arrowhead. It boasts a multiacre parking lot that provided an excellent staging area for firefighters.

chestrated attack, dropping the red fire retardant. Everybody was getting clobbered with the red stuff. A TV newscaster had it dripping from her hair as she gave her live report. It was like tomato soup on the road. A channel 4 news van was burned up, and newscaster Chuck Henry and his cameraman narrowly escaped with their lives.

The southerly Santa Ana wind kept the fire from encroaching into Lake Arrowhead, but the fire captains knew that the wind would be shifting to an onshore flow, thereby pushing the flames across the mountain. They just didn't know *when* the wind shift would occur. It changed shortly after the Santa's Village knockdown. For the time being, Lake Arrowhead was safe because the wind direction hadn't changed earlier in the day.

When I left the mountain at sundown Tuesday, it appeared that they had gotten the upper hand on "the Old Fire." At about 10 P.M. the TV reported a fire flare-up in a canyon below Running Springs. The wind was whipping it up the mountain again by Wednesday morning. The news and composition team met at my "little house" in Sky Valley for a breakfast meeting. Angela cooked breakfast while we recapped our plans for publishing and distributing these editions of the *Mountain News* and *Crestline Courier-News*. Both of our newspapers and the *Big Bear Grizzly* will be distributed "free" around the base of the mountain to all the hotels, motels and evacuation centers set up to house the thousands of evacuated families from the mountain communities.

Distribution areas will include San Bernardino, the high desert, including Hesperia, Apple Valley and Victorville and throughout the Coachella Valley, from Palm Springs to Palm Desert.

Shortly after the news team arrived at the *Hi-Desert Star* offices we watched the TV news broadcast from Lake Arrowhead. It showed cabins on fire in Cedar Glen, along Hook Creek Road. Editor Davey Porter recognized a burning house as one a few hundred feet from his home. As we went to press he learned his house was gone . . . ."

EIGHT

An Unfortunate Hunter

Southern California was already besieged by
flames on Saturday evening, October 25,
2003, when the San Diego County sheriff's
helicopter went to search for a lost hunter who
would admit later to sheriff's department investigators that he had lit a beacon fire.

According to the investigators, the hunter,
Sergio Martinez, age thirty-four, from the Los
Angeles suburb of West Covina, said he had
started the blaze so rescuers could find him.
When he was located, the hunter was said to
be dehydrated, disoriented, and remorseful and
was said to have initially blamed the fire on a
gunshot.

Helicopter pilot Deputy Dave Weldon was

sent to pick up the now infamous lost hunter. Deputy Weldon would also spot the birth of what would come to be known as the Cedar Fire. The Cedar Fire eventually killed sixteen people, burned twenty-two-hundred-plus homes, and did $400 million in damage.

Weldon would later report that at 5:45 P.M. he had requested an additional helicopter carrying a water bucket. "It wasn't growing very much at that time. It was basically contained to the hilltop," he remembered. Deputy Gene Palos said his helicopter had a bucket and was ready to go.

"As I was going over the top of Gillespie Field,* I was told by our communications center to disengage," Palos said. "We were told they would not be responding. It was too close to the cutoff time, too close to sunset."

Palos took a deep breath and shook his head.

"I don't know if I could have made a difference; anything is better than nothing," Palos added.

In classic bureaucratic fashion, Rich Hawkins of the U.S. Forest Service responded to both pilots' account of what happened. "We feel their efforts would have been futile," he said. "It was an unnecessary risk to their safety." He explained that national policies at the time dictated that all air firefighting operations had to cease a half hour before sunset.

"I admire the pilots for their courage, but we're experts in the fire service," Hawkins said. "You won't get adequate air support on the fire." Hawkins added that because the rule not to fly a half hour before sunset was a national policy there was nothing local officials could have done.

---

* Gillespie Field is a San Diego–area general aviation airport.

Meanwhile, a camera on top of Mount Laguna, about forty-five miles east of downtown San Diego, caught the early stages of the Cedar Fire.

The video shows the fire one frame every ten seconds as it worked its way through the mountains and shows the clouds of smoke growing to heights of twenty-five hundred feet. Within hours, the flames would cascade out of control and kill fifteen residents and a firefighter from Northern California between the mountains east of San Diego and the city, as well as an immigrant laborer from Mexico.

Weldon grew particularly frustrated when he saw state firefighting planes on a nearby airstrip as he approached the mountains at 110 miles per hour. He called down for help because his dispatcher had relayed reports of smoke in the area, but Weldon received no response.

A few minutes later, he spotted smoke from the fire, which at the time was only about fifty yards on each side of the small canyon and not spreading.

As he steadied his helicopter against wind gusts, Weldon's concern mounted.

Just before landing, he called for backup, asking another county helicopter—the one piloted by Deputy Palos—to speed to the scene with its 120-gallon water dump bucket. And Weldon urged the dispatcher to contact state firefighters and again renew his request for air tankers.

Weldon was again admonished that under state safety guidelines no flights were allowed to go up into the waning daylight. Cutoff on that Saturday was at 5:36 P.M. and sunset was at 6:05 P.M.

Deputy Palos's helicopter—the one carrying the dump

A FIRE SEASON BIG
ENOUGH FOR TEXAS

The term "fire season" may soon become meaningless in the western United States as global climate change potentially turns some areas into year-round tinderboxes.

On September 19, 2006, after 515 days of wildfire responses, the Texas Forest Service officially announced an end to its latest "fire season," which actually started on January 1, 2005. It was by far the worst fire season ever on record in Texas. Consecutive weather fronts in September 2006 finally dropped enough precipitation to slow fire activity to a level that could be handled within local response capabilities. Before that, the effort involved a large statewide fire response and nearly four thousand out-

bucket—flew within five miles of the fire before state officials told him to turn back. The air tankers never took off. Weldon was told crews would attack the fire in the morning.

"We were basically just offering our assistance fighting their fire, and they turned it down," said Weldon, who with his partner delivered the hunter to law enforcement officials. "I was frustrated about it, but I wasn't surprised."

The rules may help save pilots, but they were cold comfort to the son of one man who died hours after the county helicopter was called off.

of-state personnel and five hundred pieces of equipment from all fifty states.

The so-called fire "season" had lasted nearly twenty-one months. During that time, Texas Forest Service responded to 4,456 fires that burned 1,601,405 acres, or 2,500 square miles. In addition, 24,685 fires that burned 658,835 acres, or 1,030 square miles, were reported during that time to the Texas Forest Service's Online Fire Reporting System by local fire departments that participated in the system.

In all, nineteen lives were lost in Texas fires from November 2005 to September 2006. The estimated $628 million in wildfire-related losses during the latest fire season included 734 homes and 1,320 other outbuildings. Drought was widely named as the culprit in this extended fire season.

Source: Texas Forest Service, http://txforestservice.tamu.edu/.

Stephen Shacklett was killed shortly after 3:00 A.M., Sunday, October 26, when he tried to race away from the flames in his motor home. Told of that Saturday evening's events in the air, the son was incredulous.

"The hugest fire in California history," said Stephen Shacklett Jr., "and they had a chance to put it out." Though it would be of little comfort to Shacklett, the hunter Martinez would eventually be indicted the following year by a federal grand jury.

The two-count indictment would allege Martinez intentionally set the 273,246-acre Cedar Fire on U.S. Forest land after getting separated from a friend on the first day of deer hunting season.

The one positive in the entire Cedar Fire equation was that at least the authorities and the public knew how the fire started, and who the person was they believed to be the responsible party.

That was not the case for the Old Fire, and the other fires that had claimed lives and billions of dollars in lost property. The perpetrator of the Old Fire would remain a mystery, and there would be no suspects—for a while.

NINE

Simi Valley Ablaze

David Lichtman remembered he couldn't escape the smell. He likened it to the smell of a Sunday barbecue, "before you put the meat on the grill.

"The most incredible part of my commute that Tuesday was coming down the 101 toward Calabasas. There was no smoke, only clear blue sky."

For a moment, Lichtman thought the fires might be out.

The thirty-eight-year-old TV advertising executive was accustomed to spending hours of his life on the freeway, usually stopped. This day was no different. As he made his way through the bumper-to-bumper freeway traffic, he switched his CD player over to the AM dial of his radio.

"I'd never even played the AM before," he said. "Then I landed on KFI because it came in the clearest and I knew something was really wrong; they were talking about all these fires burning at the same time, and they mentioned my street in Simi."

Lichtman had just emerged from two days of meetings. In fact, he had slept the night before in his office and hadn't been watching or listening to any reports. Now, as he was stuck on the freeway, he learned his entire neighborhood was threatened. When he finally got to the divided highway that would take him through Moorpark and to Simi Valley, he met a roadblock. Nobody was allowed in; the entire area from the Ronald Reagan Freeway to the north and Moorpark to the west and south had been declared an evacuation area. The officer told Lichtman to check back later. Maybe if the fire slowed, or turned, the officer said, they might open the restricted areas to residents only.

Lichtman was frustrated. He smelled the smoke, but he hadn't seen any ash all day. Even as the day gave way to sunset, he couldn't see the glow of flames or thick black smoke anywhere.

Turning his car around and heading back toward Burbank, he tried to find a hotel room. After three tries—evidently he wasn't the only one who'd been evacuated—he gave up and decided to crash at the office. He stopped at the gym for a shower, then pulled off the 101 freeway in Studio City and grabbed a burger at Carney's. Then, like millions of Southern Californians, he flipped on the television and settled down on the office floor, hoping his home would be spared.

By 10:00 P.M. he'd drifted off to sleep, and about half an hour later his cell phone rang.

He jumped as he fumbled through his pants pockets, hoping to press the talk button before it went over to voice mail. It was his sister, Jenny, and his girlfriend, Taylor.

The women had decided to try to drive back through town on the 118 freeway. They thought since it was dark, they might have a better chance of making it around any roadblocks. They had both grown up in Simi and knew all of the side streets and back roads into the upscale Los Angeles suburb. If all else failed, Taylor knew of a little-used road that followed the train tracks through the Santa Susana Pass. They opened the windows, and the smell coming through the pass was so powerful that both women later found it impossible to describe. It was overwhelming. And then they caught their first glimpse of actual flames. There, just off to the right, coming down the mountain, was a wall of flame as far as the eye could see. They pulled over and, with Taylor's brother on the other end of the phone, cried like babies.

THE VAL VERDE FIRE: SANTA CLARITA

By now what fire officials had dubbed the Val Verde Fire had come full circle and was heading back toward Santa Clarita, which lies thirty-five miles northwest of downtown Los Angeles. In an area called Stevenson Ranch, where all the homes look exactly alike and the housing prices are staggering, self-described soccer mom Stacey Johnson was grabbing the last boxes containing irreplaceable photographs and mementos.

She called for her husband, Jim, and the couple's two school-age children, Travis, ten, and Molly, eight. Red, the family's Irish setter would cause a scare when he took off down the cul-de-sac. Jim hopped in his Lexus and cornered the dog in a neighbor's driveway. He opened the passenger-side door and Red happily jumped into the front seat, oblivious to the conflagration bearing down on them.

Jim backed the car all the way to his own house and cautioned Stacey to follow him and, whatever happened, not get separated. As they drove together out of the neighborhood, Jim found he couldn't take his eyes off his rearview mirror. In it he saw his whole life, his minivan containing the family he loved so dearly, and the telltale orange glow of what he prayed would mercifully pass his house and those of his neighbors.

Stacey and the kids didn't know it, but Jim's plan was to drive them to a motel in Anaheim and take them all to Disneyland the next day. He said as far as he was concerned, "it was in God's hands. If we were going to return later in the week to concrete foundation slabs and ashes, I figured we might as well make some good memories to counter any bad ones on the way." He smiled, thinking for the first time he understood why the emperor Nero fiddled while Rome burned.

NOTES FROM THE SIMI VALLEY FIRE

Like the fires burning out of control all across the Southland, the Internet Web logs, or "blogs" as they are commonly known, blazed all across the information superhighway. Raw emotion and quick notes jotted on a keyboard and sent over the Internet helped the writers of these entries every bit as much as the on-scene versions of events helped the readers. What follows are some of those anonymous entries:

The firefighters did a wonderful job of stamping out the fire south of 118 this morning.

The fires in Val Verde and around Piru continue to burn as separate entities with larger flare-ups happening from time to time.

The main fire north of Simi continues to rage and there are now live reports coming from the Porter Ranch area (east of the Santa Susana Pass on the LA County side of the line) tracking the eastward progress of the fire storm. It is slowly but surely creeping east.

Burning South

I'm sitting here watching the fire (NBC is focusing on Simi) that has now jumped 118 between Kuehner and Santa Susana Pass and is burning south in the area of Kuehner Drive and Los Angeles Avenue. If the fire gets loose in Box Canyon . . .

Fire Update

It's now reported that there are 10 separate fires burning in California.

I've found some additional coverage at the *Ventura County Star* (as I've been pointing out, they are doing a pretty good job with scarce resources).

With the Simi Valley fire now spilling east into LA County, the valley-based *Daily News* will probably jump into the act.

The Simi fire is now over 80,000 acres and I have heard reports putting it as high as 90,000. With all of that, the fire down south in San Diego County sounds much more out of control and dangerous. Also, various sources are reporting that the Monday Night Football game is being moved to Arizona [from Qualcomm Stadium in San Diego].

Simi Fire Update, October 26, 2003

There's not much new in the official news category. Before going to bed last night I called some friends over in the Tapo Canyon area and invited them here if they need to evacuate but they had not heard anything at that point.

The *Star* has been doing a pretty good job of getting news reports out . . . but I haven't found anyone else covering Ventura County. According to a report this morning 47,000 acres have been burned.

Sarah was out a couple hours ago and went up 23 to Tierra Rejada where she had to exit and drive west. She indicated that the fire is burning on the ranches on the south side of Tierra Rejada, which would put the fire less than a mile from the Reagan Library.

KCLU (88.3 FM) has been doing full-time coverage this morning. Someone just called in from Box Canyon and indicated that they can see fire up in the Santa Susana Pass area. There have also been some reports about a release of chlorine gas in Moorpark.

According to the latest fire update from the *Ventura County Star,* this fire burned 10 miles from somewhere outside of Piru to Simi Valley in three hours this afternoon (which explains the sudden, heavy smoke we saw).

It's been a little crazy around here this evening, with the highways closed down and much of the city driving around trying to find a good vantage point to see the fires. This sleepy little city has gone quite crazy (a weekday afternoon traffic flow on Erbes Road at 10 o'clock is one of many examples and I am quite surprised there aren't cars strewn everywhere after what we saw in a few places).

The best view Adam and I found was over at the Ronald Reagan Presidential Library. We got stopped (like everyone else) at the circular drive near the top, but parked close to the top and walked back over to the gate. We saw enough there to be scared. The fire had definitely gotten south beyond the 118 in places and appeared to be burning up the ridge just to the north of Tierra Rejada Road. There was a better view of the westernmost end of the fireline further down from the library, but we didn't stop to ogle any longer.

Sarah and I were just discussing what we might want to grab if we need to run for it. Saturday morning that would have been silly, but right now . . . who knows. If this thing manages to burn its way onto the ridges around Sunset Hills, we'll be fleeing to somewhere.

The wind seemed to be blowing almost directly south, which pushed everything in our direction. Jon wanted to go see the fire (how close I asked? about 45 feet was his reply . . . we had a little talk about the power of fire) but we passed on that. Those folks already have enough problems to deal with. I had to turn on the light in my office because I could barely see . . . 4:30 and it's already twilight.

The Piru Fire

I went out to pick up Steve from his driver's education class and on the way home we noticed a lot of smoke drifting southwest. It appears to be coming from northwest Simi . . . maybe a little further north. Will it ever stop?

As Saturday turned to Sunday, high temperatures, windy conditions, and steep terrain continued to plague the firefighters' containment efforts and smoke from the Simi Valley fires drifted as far north as Santa Barbara. The Piru Fire, which was by then burning out of control in the Los Padres National Forest also about thirty miles south of Santa Barbara, could be seen as far north as Vandenberg Air Force Base.

Residents in Santa Barbara County were grateful no fires were burning near their lovely coastal towns as of yet, but as the California Fire Chief's Association expanded a general red flag fire alert to include all of Santa Barbara County, everyone knew anything could happen. After all, the Old Fire, which began the day before, hadn't existed on Friday. Now it was threatening Lake Arrowhead and the mountain resort communities from Crestline to Big Bear.

The red flag alert was initially issued Friday but only included the south coast area between Gaviota and the Ventura County line. "Local fire agencies have staffed additional firefighters in anticipation of hotter and dryer conditions accompanied by strong winds during the next few days," the alert stated.

In addition to more personnel being added to wildfire engine crews, the alert announced the cancellation of firefighter vacations and the callback of off-duty employees to make up for the personnel working on fires outside the county. While Santa Barbara experienced the warm and dry temperatures, it thankfully missed the wind that fanned the flames in Simi Valley and elsewhere around Southern California.

An average of fifty-five to sixty firefighters from departments within Santa Barbara County were deployed as part of the interagency strike teams that were sent to aid firefighting efforts in Ventura and Los Angeles counties, and area fire chiefs employed twice-daily conference calls to keep everyone updated on the situation.

The local 911 emergency dispatch centers were inundated with calls from people living along the Santa Barbara coastline wondering about the drift smoke, which spread along the coastal Santa Ynez Mountains and into the Santa Barbara Channel.

As part of the red flag fire alert that critical weekend, the local fire departments requested the Amateur Radio Emergency Services groups to patrol roads in areas of severe fire danger. In addition, the Santa Barbara County Sheriff's Department expanded its aerial coverage to spot new fires as quickly as possible.

As of Sunday morning, the Piru Fire, which was by then five miles northeast of the community of Piru, had burned 1,253 acres. However, by Sunday afternoon, Forest Service Fire Information Officer Joe Pasinato said the fire had swelled to an amazing ten thousand acres. By 8:00 P.M., the fire had consumed nearly twenty-five thousand acres—an area of almost forty square miles—and the Ventura County

Sheriff's Department responded by recommending voluntary evacuations of Piru and the nearby town of Fillmore.

The 784 firefighters and support personnel who were
working diligently to contain the flames had experienced a
notable setback.

Earlier that Sunday morning, the U.S. Forest Service had
estimated the Piru fire to be 85 percent contained, but the
number had dropped to 5 percent late the same night. A total
of six firefighters sustained minor injuries while battling the
blaze, but all were treated and released back to their camps.

THE SIMI VALLEY FIRE

As the Simi Valley Fire continued its move toward Stevenson
Ranch, the eastern flank extended from just west of Interstate
5 near Six Flags Magic Mountain and south to Chatsworth.
The topography and the dense fuel of the area were aiding
the flames.

The Simi Valley Fire had actually started northwest of
Santa Clarita, in Val Verde, and spread to Ventura County
and then turned and began to burn back to Los Angeles
County.

One Super Scooper firefighting aircraft and three county
helicopters, as well as other fixed- and rotary-wing aircraft
from Los Angeles County and other agencies, were assigned
to the blaze. Bulldozer and hand crews cut new fire roads and
set backfires to try to keep more homes from burning.

One particularly gruesome TV image showed a wall of
flame descending on one of the restored homes at historic

Mentryville, the old California oil and gold town. But fire-fighter Roland Sprewell told the channel 4 newscaster his crews purposely started the fire just behind the house to "backfire out" some of the brush at an opportune time. He said the operation was successful. But farther west, more than two thousand homes in Simi Valley were still in danger when a smaller fire jumped State Route 126. Firefighters were also working to save the Ronald Reagan Presidential Library in the city.

By noon on Monday, October 27, Ventura County fire officials confirmed the loss of twelve homes and said the Union Pacific Railroad had closed all rail lines into Simi Valley in the southeast corner of the county, bordering the San Fernando Valley of Los Angeles.

Flames also came to within a quarter mile of the Federal Aviation Administration's radar facility station. Air traffic controllers transferred their responsibilities to a facility in Palm-dale, and the switch delayed air travel for several hours at several Southern California airports, including Los Angeles International and San Diego International.

At 3:36 on Monday afternoon, firefighters took advantage of the lull in the winds, however brief, to put helicopters and fixed-wing tanker aircraft back into the air to dump water and retardant on the fires. They had been grounded since sundown on Sunday.

Meanwhile, hundreds of evacuated residents of the heavily populated suburbs waited in their cars, on the streets, or at shelters for word on the fate of their homes. Among those sitting in their vehicles watching the burning skyline that day were Sharon Robinson and her daughter, Kim, who had fled their home after throwing whatever clothes and other belongings they could into the back of their truck.

"We've lived in our home for thirty-five years," Sharon Robinson recalled. "Fire had always stopped in the foothills before. I never thought it would reach our home."

"I looked outside my house and I thought I was in the middle of hell; it was redness everywhere, unbelievable," recalled Joe Wronowicz, who, along with his family, put off evacuating, putting their faith in firefighters to protect their Rancho Cucamonga neighborhood. That morning Governor Davis asked President Bush to declare disaster areas in the four counties affected by the fires, an act that would, it was hoped, pave the way for financial aid.

"My heart goes out to them," Davis would say at a news conference in San Bernardino, describing those who had lost or been forced to abandon their homes.

"This is a terrible situation. They are the worst fires in California in 10 years," Davis said. The governor noted he had authorized more than 650 fire engines to help the effort.

IT'S LIKE A WAR zone down here, Fire Engineer Greg Leptich thought as he battled the flames of the Simi Valley Fire. "I've been to a lot of fires, but nothing like this." By Thursday, October 30, the fires in Southern California had raged through 750,000 acres, an area more than three-quarters the size of Rhode Island. As Leptich continued to fight the blaze with his team the tally was coming in: There were twenty fatalities, including one firefighter, more than twenty-six hundred structures were lost, and the firefighting costs were well into the billions of dollars. *By the time this is over,* he recalled thinking during a short break from the Simi Valley Fire line, *this will be the most extreme and most expensive fire in California history.* Leptich was almost right.

TEN

The Belly of the Beast

The Cedar Fire was reported on Saturday, October 25, 2003, at around 5:30 in the afternoon. This would be the fire that eventually would become the largest single wildland fire in California history to date and would consume 280,278 acres, destroy 2,232 structures, twenty-two commercial buildings and 566 outbuildings, and damage another 53 structures and 10 more outbuildings before it would end its gluttonous rage.

It would also take the lives of fifteen civilians, would injure 107 more, and would take the life of its staunchest enemy, a firefighter.

Steven Rucker, age thirty-eight, a beloved husband, son, and father, literally laid his life on the line for others and would pay the ultimate price.

At approximately 7:00 P.M. on the evening of October 27, the Novato Fire District, just north of San Francisco, received a request from the Marin County Emergency Communications Center for a type III engine to respond to San Diego County to assist with the fire there.

A four-person crew consisting of Captain Doug McDonald, Engineer Shawn Kreps, Engineer Rucker, and firefighter/paramedic Barrett Smith was assigned from an established list to staff Engine 6162.

McDonald's crew had just returned to work at 7:30 that morning after having enjoyed a four-day break. They had not worked any extra shifts during the preceding few days. They also had what the investigation analysis would later term as a light to moderate call volume that day, and prior to the actual dispatch the crew spent time collecting the items they would need for their extended assignment.

The communications center would officially dispatch Engine 6162 at 9:27 that evening and the crew would depart Novato Fire Station No. 4, heading to the town of Ramona in northern San Diego County.

They traveled nonstop to the truck scales at Livermore on eastbound Interstate 580 and once there met up with the task force leader and assistant task force leader from Lawrence Livermore Lab and two of the engines—Engine 1541 from Camp Parks and Engine 334 from San Ramon Fire Protection District—that would eventually compose the five-engine Task Force XAL2005A.

The group was designated a task force because it contained a mixture of type I (structural) and type III (wildland) engines.

The group left the scales at midnight and joined another convoy in Santa Nella along Interstate 5 in the middle of the San Joaquin Valley. Every hundred miles the crew would rotate drivers.

Upon their arrival at Gillespie Field, the airport in El Cajon, just east of San Diego, the crew managed to get a couple of hours of rest before reporting for structure protection duty at the intersection of Interstate 8 and Highway 79 in Lakeside, about five miles northeast of the airport.

Their assignment was Riverview Road, designated "Camp Oliver," where they supported a burn-out operation, in which backfires are set. They were instructed to hold in the area until 9:30 P.M. and patrol for hot spots.

During their operations on Riverview Road, the crew encountered moderate fire behavior as the fire made downhill runs in heavy brush.

The crew returned to the Gillespie Field staging area at 10:00 P.M. for dinner and then bedded down for the night. Engine 6162 members recall going to sleep between 10:30 P.M. and midnight. They woke between 5:00 and 6:30 A.M. on October 29.

The crews were rested and in good spirits. Their task force leader attended the morning briefing at approximately 7:00 A.M. while the crew ate breakfast and completed their typical morning checks on the engine. They ran the pump, completed the pre-trip brake check, and conducted a check for operational readiness.

The task force leader briefed the group on its assignment and shared safety information. A fifth engine, Engine 71 from Ramona, was added to the task force.

Rucker turned in a defective radio to Communications and picked up a replacement so there would be two high-band portables for the four personnel on Engine 6162. One of the radios would later be given to the task force leader because he didn't have one.

At 8:30 A.M. Task Force XAL2005A left the staging area and headed for the assignment. They arrived at Santa Ysabel, east of Ramona in north-central San Diego County, at about 9:15 A.M.

At 9:47 A.M., Captain McDonald and the task force observed that the fire had crossed Highway 78/79 between Santa Ysabel and the Inaja firefighter memorial to the west. According to McDonald, he could see engines backing down the highway to avoid fire that had crossed the highway. By about 10:20 A.M. the task force was assigned to work structure protection in the Riverwood Estates. McDonald was concerned about passing through the area where the fire had recently crossed Highway 78/79; however, they managed to navigate the area without incident.

The 6162 crew noted that when they arrived in the Riverwood Estates a firing operation had already been conducted around the structures, and they started mop-up operations. They watched as the fire burned past them to the west in the distance, the area northwest of Highway 78/79 where the fire had jumped the highway. In fact, at that point the fire was calm in the Riverwood Estates area.

While Captain McDonald and his crew were doing structure protection in the Riverwood Estates, he talked to his crew, including Rucker, about the use of hose packs.

McDonald said that if there was a need to protect a

structure, they would add to the one-hundred-foot-deployed hose lines an additional hundred-foot length of hose from one of their two-hundred-foot wildland hose packs. At about 11:00 A.M. the task force leader met with Division I and learned he needed to move his task force to Orchard Lane because of an increase in fire behavior in that area requiring the apparent need for immediate structure protection. There would be no specific briefing of the task force at this time.

At about 11:45 A.M. the task force leader drove north along Orchard Lane while the task force staged at the intersection of Orchard Lane and Highway 78/79. He called back for the Task Force Team to start moving up, and at this time McDonald and his crew observed the fire crossing Orchard Lane at the north end. McDonald's Engine 6162 was the last engine in line, and at about 11:50 A.M. they arrived near 915 Orchard Lane. They immediately began scouting structure locations near the north end of the lane.

Due to the fire activity, McDonald ordered his engine backed up to a field identified as a safety zone. The other units backed up with them.

Engine 6162 was assigned to 920 Orchard Lane, the second home from the north. Captain McDonald and the task force leader discussed the placement of McDonald's engine, the need to deploy hoses, and to clear the area. McDonald and his crew were comfortable with the plan. He ordered Shawn Kreps to back the engine up the cement driveway at 920 Orchard Lane and walked up the driveway ahead of the engine to evaluate the location.

Firefighter Smith and Engineer Rucker cut brush hanging

over the sides of the cement driveway and directed the backing of the engine. Having walked only as far as the residence's detached garage, McDonald returned to his engine. After assessing the situation, he expressed some concern about the conditions to his crew. McDonald radioed the task force leader and advised him that he didn't think the house was defendable, so the leader told them to come out.

McDonald, Rucker, and Smith walked to the end of the driveway to assess the conditions closer to the single-story stucco home with a composite roof. As they continued past the front of the house, which faced west toward the San Diego River drainage ditch and well beyond where McDonald stopped on his first assessment of the area, they surveyed their locations and found the brush had been cleared away below the residence on the west side for some 150 feet.

Because of the conditions they now observed, they decided the structure could be defended. Since the brush to the west had been cleared, there was a view across the canyon to the west and northwest. Tall brush and drifting smoke restricted the view to the southwest.

From their location, the 6162 crew saw smoke to the north and determined it to be the main fire and the primary threat. They predicted this fire would continue backing down the ridge to their location and that it would remain what is known as a flanking fire (fire that is moving perpendicular to the wind).

Small runs of fire were occurring across the canyon on the west side of the drainage as the main fire backed down into the bottom of the canyon. No fire activity was visible to the southwest. There was a wind blowing up the canyon and upslope toward the location of 920 Orchard Lane. The wind speeds were estimated at 7 to 10 miles per hour.

The crew wasn't aware of any other crews operating on Orchard Lane other than those assigned to Task Force 2005A. The crew was also not aware of the exact location of the other resources assigned to the task force.

After Engine 6162 arrived at the top of the driveway, Mc-Donald conducted a briefing, with the crew explaining the plan of operations, structure preparation, safety, and survival. An ax was placed at the rear door of the structure in preparation for forcible entry, should it be needed. The chain saw was placed on a cement patio on the south end of the house after the crew had used it for additional brush clearing. The roof was laddered near the front door using a ladder found at the residence. The crew then charged up the front bumper line with water to be used as an engine protection line. Two hundred-foot-long, 1.5-inch hose lines were deployed and charged to protect the structure. These two hose lines were "wyed off" a rear discharge on the engine.

The crew did additional work preparing the residence for the oncoming flanking fire from the north, including the removal of additional brush downslope from the engine. They moved the brush away and backed the engine up about ten feet to avoid subjecting it to heat from a brush pile that they lit below the driveway.

Captain McDonald walked north along the driveway, and while nearing the front door of the house he saw a CDF pickup truck drive up to the top of the driveway and park near the garage. The driver was a CDF captain who, he believed, was preparing to conduct a firing operation near the garage.

He was not able to speak with him before he was out of sight, but McDonald thought the CDF captain saw him.

McDonald met with his crew and told them about the CDF captain and the firing operation. He then instructed his crew to fire out the stubble grass area below the driveway on the west side. Kreps walked down the driveway toward the garage area and observed the fire already on the ground behind the garage and to the north in the brush. Kreps believed this was the main fire backing down the ridge as expected. He began his strip firing as Captain McDonald threw fusees (small fire starters) down the slope into the heavy brush. None of the crew members noted any effect from the use of the fusees, and none of them was aware of the exact time the CDF pickup left their location.

At about 12:35 P.M. the task force leader and his assistant arrived to check on 6162's progress and plans. The sky was clear overhead and the winds were moderate, still upcanyon and upslope. They spoke with McDonald, the Task Force Leader telling the men what a good job the 6162 crew was doing. They noted the main fire was continuing to back down the ridge from the north and was still about three hundred yards north of their location.

About five to eight minutes after the task force leader left the scene, the crew of 6162 observed an increase in the fire activity downslope from their location. The crew gathered at the engine to discuss the situation, and as the fire intensity increased they decided to move to the passenger side of the engine away from the radiant heat below them.

At this point, Smith staffed a hose line at the front passenger side while Rucker staffed a similar hose line near the rear passenger side. Kreps was standing at the rear duals on the passenger side with his back to the engine. Captain McDonald remembers being just to the rear of Smith.

Kreps would later recall all four members of the crew made it around to the passenger side of the engine and that Rucker was directly in front of him, no more than five feet away. McDonald had the only portable radio.

As Smith took his position at the front of the engine, he noticed hot embers blowing into the juniper bushes on the patio behind him. He used his hose to knock down some fire in the bushes.

As Kreps stepped around the rear of the engine to snap a photograph, Rucker shouted, "Get back!" According to Kreps, in the previous twenty seconds conditions were stable; however, that began to change dramatically.

Fire activity at ground level as well as the smoke appeared to be drawn to the south or southwest. Smith describes the conditions as changing from full sunlight when they went around the engine twenty seconds before to twilight and then darkness from smoke and the orange glow from fire embers blowing past the engine.

The heat intensity increased dramatically.

Kreps remembers, "This is when the sky got dark, it started getting very hot, embers were flying everywhere, and it sounded like a freight train coming."

Captain McDonald remembers at about this time the sky started to turn orange and the conditions began to be untenable. According to McDonald and Kreps, the majority of the heat was coming from the south or southwest. McDonald recalls seeing a huge orange glow as he looked back over the top of the rear of the engine. He could see the orange glow through the trees and the tall brush. Kreps kept his mouth covered with his gloved hand and reminded himself to take shallow breaths to protect his airway.

They all noticed a major wind increase and then Smith saw the flaming front, blowing across the driveway in the direction of the garage, cutting off their egress to the safety zone in the field off Orchard Lane. The juniper and boxwood bushes planted along the cement patio behind them suddenly burst into flames. The situation was deteriorating rapidly and the crew decided they should take refuge in the structure. Captain McDonald ordered his crew into the house.

THE OLD WATERMAN CANYON FIRE RAGES ON

Dozens of firefighters manned a battle line along Highway 18 in a last-ditch effort to prevent the Old Fire from racing into Crestline, where, fueled by tens of thousands of dead pine trees, it had the potential of exploding into an unstoppable firestorm. The fire, already having consumed twenty-six thousand acres and hundreds of homes below, snaked up toward the Rim of the World Highway, and firefighters took a stand near Lake Gregory Drive to prevent that conservative 26,000 figure from climbing. USFS Captain Scott Howes was adamant. "Remove the fuel ahead of the main fire!" he yelled.

"We're trying to starve the fire by removing the fuel that's just ahead of it!" Crest Forest Fire Protection District Chief William Bagnell barked into his radio.

"Forget that! You gotta get the *main* fire fuel gone. Now!" Howes yelled.

Bagnell wasn't accustomed to such disrespect, especially from a Forest Service captain and especially in his own district. He turned to Lance Snider, a five-year Crest Forest firefighter and a popular member of his crew.

"Whataya think?"

"I think they don't know Crestline."

"Is it worth getting the Hotshots in here to do what he says?"

Snider took a breath and looked around at the orange glow and noted the sounds of the encroaching fire. "Hear that?" Snider said, wanting to get back to work.

"I hear," the chief exhaled. "Do it."

Within moments the team shifted position and moved toward the advancing wall of flames. They would remove the fuel, just as the Forest Service captain had suggested. While Bagnell didn't exactly appreciate being shown up by another agency, that didn't matter now. The only thing on his mind at the moment was beating down this horrid beast that had already claimed twenty homes in his town. One of the homes belonged to one of his best friends, and Bagnell was in no mood to see anyone else in his town lose a home.

While the fire was snaking north, sheriff's investigators released a composite sketch of a man suspected of starting the fire. Since Saturday, at least $225 million in damages had been racked up in the Old Fire, and it had already caused the deaths of two men. Investigators were seeking a white man in his early or mid-twenties that was seen driving a light gray van away from the fire's point of origin.

Near Crestline, swirling thick smoke and ash enveloped more than fifty firefighters as they shot water into the flames dancing over their heads and along the edge of the highway. With the Santa Ana winds still howling, they had three immediate objectives: protect Crestline, Rimforest, and Lake Arrowhead.

As the fire stormed across the foothills, residents in the northeast section of East Highlands Ranch were ordered to evacuate. The fire, which appeared to have calmed earlier in the day, revved up and was spreading rapidly late in the night. But when the blaze coming toward them was extinguished, residents relaxed.

Earlier in the day, thin, snakelike belts of fire moved down the hills above the neighborhood. A few folks in the neighborhoods nearest the hills hung close together at Pleasant View and Van Leuven to watch the smoke billowing above them but heading northeast. Some had been there since 3:00 A.M. Many did not go to work. Clusters of people stood calm and curious, more in awe than in panic. Ash descended steadily as towers of smoke rose, but residents saw no reason to leave. Late Monday, they had plenty of reason to leave as the blaze came over the mountain and threatened the homes in the area. In an effort to cut the fire off, firefighters had set backfires.

The fire crew working the rim high up on the mountain had the same idea. Set the backfires and perhaps avoid letting the fire burn all the way to Big Bear. The method had already saved the Rim of the World High School from destruction on Tuesday; why wouldn't it work again?

Fireman Dean Hansen took his gel torch from the engine

and crossed the highway to the east of Kuffel Canyon. The flames took hold, and the firefighters repositioned their engines to a point about a quarter mile ahead of the flames, readying their hoses in anticipation of the flames that would ultimately burn a fifty-yard-wide swath before the hoses would even be turned on and the helpful backfire would be doused.

Then came the unthinkable: The wind switched, sending embers across the disposal site at Heaps Peak.

By 1:45 P.M. the embers grew into a wall of fire that fully engulfed Cedar Glen, and by 10 o'clock on Wednesday evening more than three hundred homes would be gone.

"THERE'S SOME BAD NEWS," *Mountain News* publisher Harry Bradley cautioned as I walked the darkened hall to *Hi-Desert Star* editor Stacey Moore's office in Yucca Valley, northeast of the San Bernardino Mountains. Our *Mountain News* staff had taken over the place, the computers strung together in a haphazard fashion along banquet tables and unused desks. Moore's office had the only television, and I wanted to keep up with whatever unedited images the Los Angeles TV stations were throwing on the air. "What now?" I asked. Then my boss gave me news I didn't want to hear.

Fires and evacuations are one thing, but none of us was ready when he told us our colleague Ken Carriero, had passed away at St. Bernardine Medical Center earlier in the day. Ken was gone. And I didn't have time to give it another thought. My mentor at the *Mountain News,* a guy we all loved, the kind of guy who'd forgotten more than any of us would ever learn in a lifetime, had left this world.

The gallows humor began almost immediately, me guilty for a lot of it. "Boy, Ken sure went out in a blaze of glory!" "You know, Ken was always worried about getting fired!" "Hey, maybe ol' Ken'll take one of his famously long diabetic pisses and put out the fires for us!"

A newsroom can be a pretty sick place when it comes to humor. Just know that we knew Ken would be laughing the loudest.

Then I turned the television on.

Channel 5 had the most reliable coverage, but they tended to focus more on the Simi Fire, so I flipped over to KNBC anchorman Paul Moyer and hoped I'd catch some shots of the mountain. Moyer and crew didn't let me down. There was my house—I was certain of it when I recognized our one-of-a-kind fireplace—burning to the ground.

Karen called. She had been crying.

"Babe, I think I saw our house."

She had. And we were minutes away from deadline.

"Sweetie, can I call you back? I'm gonna check on something and make sure it's our house, OK?" Then I hung up and pulled up the NewsEdit document on my computer screen. I had to update the story on the destruction in Cedar Glen to include the block of homes where my neighbors and I once lived. We hadn't even gone to press and I was already referring to my home in the past tense. It's how I kept my sanity. That and knowing that we had a job to do for all our friends and neighbors parked at the evacuation centers in San Bernardino and Hesperia. That, and not knowing what else to do.

By the next morning, Thursday, October 30, 2003, the

word came down that no one would be allowed back up the mountain for at least a week, maybe more.

Following the very disappointing announcement, Harry arrived with some good news. He said the paper was going to spring for some private resort trailers near Palm Springs and that we would be given an expense account until things got back to normal. That meant I would get to be with Karen and the kids, the first step toward healing a very gaping wound.

Within the space of a few hours I had lost a mentor and friend, as well as my house, and we didn't even know what the status of the newspaper would be. Advertisers were canceling in droves, and who could blame them? Rumors were rife that the entire town of Crestline had burned down, that Cedar Glen had lost its business district as well as the malt shop, and that by the time it was all over there would be nothing left to go back to. What good is a newspaper if there are no homes to deliver it to or customers to read it?

I drove the long detour route to Orange County in the company truck. As I arrived at my parents' place, the eerie sights and smells of fire and smoke were replaced by the gaiety and revelry of shoppers, diners, and moviegoers visiting trendy Downtown Brea. What was wrong with those idiots? Didn't they know people were losing their homes and livelihoods? Didn't they care?

No.

I went inside and found my family waiting, every piece of furniture in my mother's living room occupied by someone I loved. Gracie, our four-year-old, was oblivious, lost in the Disney Channel, which was where I'd hoped she'd be. I wished I could be there as well.

"Well," I uttered, "if we're gonna be homeless, we might as well do it together."

Karen met me with a hug, and she wouldn't let go. She just cried. The older kids' faces were drawn, and they tried to process as much as their teenage brains would allow. They chose to deal with our loss by walking to Tower Records and buying CDs. I'm convinced if my kids had been aboard the *Titanic* that fateful night they would gladly have exchanged a life jacket for headphones and a CD collection of the latest devil music, God love 'em.

We packed up the van and the company truck in preparation for the move to the trailer resort. The next morning we said our good-byes to my folks and headed back to hell. We didn't know exactly where we were going, but we were going together.

ELEVEN

"One of Our Own . . ."

CEDAR FIRE INCIDENT: ENGINE 6162, TASK FORCE XAL2005A

Barrett Smith, responding to Captain Mc-Donald's order to get into the structure, immediately dropped his line and ran in the direction of the three-foot-high elevated patio. Once he left the protection of the engine, Smith experienced severe thermal conditions. He leaped past the burning bushes and onto the patio. He was followed by Kreps, who ran to the steps, stumbled, and fell to his knees at the top of the steps but then recovered and continued to retreat to the rear of the house following Smith. They both covered more than fifty yards in their run from the engine to the back door of the structure.

McDonald moved toward the steps of the patio and accounted for Kreps and Smith while they ran onto the patio and around the rear corner of the house. He then turned toward the back of the engine to look for Rucker, not understanding why Engineer Rucker was taking so long. The captain located Rucker standing where Kreps had last seen him. As McDonald approached, he noticed Rucker appeared to be disoriented.

McDonald began to yell for Rucker to move to the structure. Rucker was slow to respond, then turned his head and almost started walking the opposite way. McDonald continued to yell at Rucker to move toward the structure, and Rucker responded by moving toward McDonald, who recalls having to cup his hands to his mouth in order to be heard over the roar of the fire.

McDonald saw Rucker take two steps toward him and fall to the ground and then stand up, on his own. Rucker then turned toward the tailboard of the engine and, according to McDonald, "he might've taken a half step in that direction."

"I remember he looked in a southerly direction at the approaching wall of fire. He stepped toward me and then turned toward the bushes along the patio and fell face-forward into the burning bushes." Captain McDonald moved immediately toward Rucker, assisting him up the steps leading to the patio. Rucker was able to climb the steps under his own power, "although," McDonald would recall, "he was slightly hunched over at the waist."

According to Captain McDonald, sometime between

Engineer Rucker's first fall and when they reached the steps to the patio, Rucker said, "I'm burning up!" But McDonald couldn't see any fire on Rucker.

The captain followed Rucker up onto the concrete patio, where Rucker fell the final time: first to his knees, and then face-first onto the patio without any attempt to brace his fall. Engineer Rucker wasn't responsive as McDonald unsuccessfully tried to get him onto his feet.

McDonald immediately issued a "firefighter down" call on his portable radio and then recounts dropping the radio after he made the call because it became too hot to hold. He recalls turning a full 180 degrees and trying to pull Rucker up. McDonald yelled for the other two crew members to return but realized he couldn't be heard due to the noise of the fire. When he realized he was being burned and that his friend, Rucker, was beyond help, McDonald made his way around the rear corner of the house.

Meanwhile, Smith and Kreps were at the back of the house, where they used the ax and multiple kicks to force the door open. They both entered the residence and, turning around, realized they were alone. They returned to look for Captain McDonald and Engineer Rucker. Smith and Kreps traveled along the back of the house to the south end, using the house as protection. As they arrived at the corner, McDonald staggered out of the flames, burned and dazed.

The captain told them Rucker had fallen and they needed to go back for him. McDonald himself, in spite of his burn injuries, went with them to find and retrieve their fallen comrade. Based on their observations of the conditions, Kreps and Smith determined the patio area was not defensible or safe. Taking McDonald with them, the three retreated to the rear door of the residence.

Once inside, McDonald realized he was burning and took off what remained of his web gear. He searched for his portable radio to call for help, but he forgot he no longer had it with him. The three men discussed an alternate plan to reach Engineer Rucker.

After a moment, the decision was made to open the front door slowly and check to determine if the outside of the house was tenable, but intense heat surged through the small opening and they quickly closed the door. After another few minutes, they decided to check the front door again, and this time the heat had subsided.

Kreps exited the doorway in search of his missing friend, as Smith attempted to exit the door but was forced back by a wave of heat. Smith remained inside the house, caring for McDonald.

Kreps moved toward the front bumper line of the engine, which was still running at a high idle, just as he had left it. He took small, shallow breaths and used his shroud to protect his airway and as he passed the front of the concrete, he saw the body of Engineer Rucker lying midway across the patio. Kreps continued to the bumper hose line and advanced it toward Rucker. He opened the nozzle but only had ten to fifteen seconds of water—the fire had burned through the rear hose lines on the engine, and the tank had been pumped dry. The engine, however, appeared to be basically undamaged and drivable. They were running out of time.

· · ·

IN AN INSTANT, THE fire intensity increased, forcing Kreps to take shelter in the backseat of the engine. Once inside he continued deploying the extra fire shelters stored in the cab. Knowing there was nothing he could do for Rucker and concerned that Smith and McDonald would come searching for him, Kreps took a single breath and ran back to the front door of the house. Smith, from inside the house, opened the door intending to search for Kreps, who at that moment burst through the door.

While the three of them sat inside near the front entrance of the house, the north end of the structure began to burn. Smoke banked down to waist level, dropping like a curtain, and the trio decided to attempt a return to the engine. Knowing there was no hope for their friend Rucker and that the house would soon be fully involved, the three exited the front door and headed for the safety of the engine. Captain McDonald was helped into the front passenger seat as Smith ran to the rear of the engine and disconnected the two protection lines. He climbed into the rear passenger seat while Kreps mounted the driver's seat.

As Kreps began driving the engine slowly down the driveway on a northerly path toward Orchard Lane, heavy dark smoke obscured the driveway and he was forced to "feel" his way down the concrete, using the feel of the tires as they dropped off the edge in order to gauge corrections.

At one point he stopped the engine to avoid running off the road and getting lodged against a tree. Kreps knew there was a significant curve in the driveway, which, if he missed it, would take them off the path and down a steep slope. The crew was still concerned about being overrun by the fire once again. A gust cleared enough smoke for Kreps to make an adjustment, and they were able to move forward.

During the descent of the driveway, Captain McDonald was able to transmit a "firefighter down" message on Command Net using the mobile radio.

Kreps drove to the bottom of the driveway, onto Orchard Lane, and headed south. McDonald made another announcement on the Command Net and, disoriented by the smoke and advancing flames, ordered the engine to stop in the road and got out. The crew members immediately helped him back into the engine and continued toward Highway 78/79.

At about 1:10 P.M. the three members of the 6162 crew arrived at the south end of Orchard Lane at a location just short of the highway where they found the Plumas Hotshot Crew, who feared some of them may have been burned. At that time the advanced life support (ALS) ambulances and medical helicopters were requested.

Captain McDonald, in need of immediate ALS intervention, was triaged off the fire line as the most severely injured, with first-, second-, and third-degree burns, as well as burn damage to his airway. Prior to being treated for their own burns, Kreps and Smith, the only paramedics at the location, used the ALS equipment from their engine to start two IVs on their captain.

At about 1:25 P.M., the first Julian paramedic ambulance arrived at Engine 6162's location.

Kreps insisted McDonald be transported in an ALS helicopter with a nurse so he could be chemically paralyzed—

thereby rendering him unable to further damage his burned skin by rolling on the gurney—and then provided a stable airway through a process known as rapid sequence intubation (RSI) prior to arriving at the burn center. McDonald was packaged and transported to a landing zone where CDF Helicopter 202 met him at 1:52 P.M. Copter 202 then transported McDonald to Ramona Airport to rendezvous with a Mercy Air helicopter. Captain McDonald was transferred to the Mercy Air crew, who intubated him and flew him to the San Diego Burn Center.

At about the same time, CDF Copter 406 took Kreps and Smith to the Ramona Airport, where they caught another Mercy Air flight to the same burn center. In the space of two hours, lives had been changed forever.

THE CEDAR FIRE: JULIAN

The cries could be heard over the roar of the flames. "It just swept right over them. They probably didn't have time to get out of the way," San Diego County Sheriff's Sergeant Conrad Grayson said. "I was hoping we wouldn't have to do this with a firefighter or a deputy." Grayson was talking to a local reporter as one hundred fire engines encircled the historic mining town of Julian in the mountains of eastern San Diego County.

Saving the town of thirty-five hundred, a popular weekend getaway in the mountains of eastern San Diego County, renowned for its vineyards and apple orchards, was the county's top priority, especially since the firefighters had learned they had just lost one of their own.

But as the winds picked up, floating embers sparked spot fires near town, forcing some crews to retreat. Some two dozen engines and water tenders that were headed to Julian were forced to turn back when flames swept over Highway 78, just east of Santa Ysabel. South of Julian, about 90 percent of the homes had been destroyed in Cuyamaca, a lakeside town of about 160 residents. Charred cows lay by the side of the road, and stone entryways stood in front of houses that were reduced to mountains of rubble. Bill Bourbeau, a safety officer for the Cleveland National Forest, shouted a general warning into his walkie-talkie.

"Everything's kind of happening all at once. These fires are trying really hard to tie in with each other." Bourbeau was with a crew along Highway 78. "It's tremendous."

San Diego County fire officials worried for days that the 233,000-acre Cedar Fire and the 50,000-acre Paradise Fire would merge into a huge single blaze that would make it nearly impossible to keep the flames from reaching Julian.

"There are ranches and little communities which make it really difficult to fight a fire like this. . . . It was almost overwhelming," Bourbeau would later recall. "It was so big, and we're still trying to get a handle on the organization part of it. What a joke."

As Bourbeau and crew carried on, the next civilian victim was found Wednesday at a home in Alpine, consumed by the Cedar Fire. A woman who lived at the address had been reported missing, and officials in the county predicted the death toll would rise after investigators began scouring devastated neighborhoods.

On Wednesday, a crew of USFS Hotshots directed to cut fire lines around Julian was given an ominous warning by their team leader: If they came across any human remains, they were told to cordon off the area until a medical examiner could get in later. Firefighters were battling westerly winds sweeping inland from the Pacific. The cool, moist breezes replaced the hotter and drier Santa Ana winds that had whipped fires into raging infernos over the weekend, but also confounded firefighters by directing flames toward mountain communities.

THE CEDAR FIRE: THE AFTERMATH

In the weeks following the Cedar Fire, the San Diego County Medical Examiner's Office was able to provide details about how firefighter Steven Rucker was killed while battling flames at a home near Julian.

It appears Rucker may have been lugging a chain saw and other tools to save the home on Orchard Lane in the tiny town of Wynola when, while heading for shelter, he tripped and was overrun by the racing flames.

The coroner's report said only the sole of Rucker's right foot was still intact after the flames finally died down and his crewmates arrived. Rucker's radio, chain saw, and tools and an unopened emergency fire shelter were found beside his body.

Though he was critically burned, Captain Doug McDonald survived. The other two in the crew suffered lesser injuries.

Half of Scripps Ranch, an upscale inland neighborhood in the northern part of the city of San Diego, was evacuated as fire threatened thousands of homes, and when the fire was finally over 343 homes in that area were destroyed and many lives changed, including those of the firefighters at the scene.

When Fire Captain Dan Saner visited his home the first time since the fire hit his neighborhood he recalled, "Smoke was down to our ankles. You could hardly see. It was one of the most amazing things I've seen in my life.

"When you turned and looked into fire coming over the hill, hit with golf-ball-sized embers, it was difficult to even be out here."

Four firefighters, including Saner, battled the fire and managed to save some homes.

"The wind was blowing at sixty miles per hour," Saner said. "The only thing we could do was stand in the way of part of it. I'm just happy to see homes standing. It is great to see that some of our work really paid off."

According to the copious notes kept by a firefighter, Colin Wilson, who was on the front lines of what would come to be known as the largest fire in the history of California, "The Cedar Fire was not in any way a typical incident."

Wilson, a firefighter with the Anderson Valley (Northern California) Boonville Station, described in a report he presented months later how the conflagration strained the considerable resources of the State of California to the breaking point and beyond.

"Usual and customary procedures and rules sometimes went out the window," Wilson reported, "not by design, but because way too much was happening in way too many places and way too fast." Wilson said the Incident Command System was in some cases unable to grow fast enough to meet requirements on the fire line.

"The incident was also managed by multiple agencies under a unified command, which also complicated things," Wilson declared, "which to my mind [meant] there just wasn't a good answer to these problems."

According to Wilson, large disasters are by their nature overwhelming events that resist the best efforts to control them. Wilson emphasized in his report that "Eventually you get to the backside of the thermal curve, the fires die down, the winds abate, the waters recede. The trick is to just get through it, event by event, protect yourselves and your equipment, and, when you get the chance, make things better one small piece at a time."

A HERO COMES HOME

At his home firehouse in Novato on Thursday, October 30, Engineer Steve Rucker was remembered as a hard worker dedicated to the job.

Doing his best to hold back the tears, Novato Deputy Fire Chief Dan Northern said, "He wasn't sent there. He asked to go."

Fred Batchelor, a state fire marshal, said the crew was overrun so quickly they didn't have time to reach their engine. He recounted how the crew of Engine 6162 tried to find refuge in the house they were trying to protect.

Rucker's fellow firefighters remembered the eleven-year

A LETTER FROM THE RUCKER FAMILY

It is a testament to the family of Steven Rucker that even during the darkest days surrounding his funeral, they would take the time to issue this statement to his brothers and sisters on the fire line.

November 11, 2003

We always knew Steve was special. Your outpouring of support has been overwhelming. We are truly humbled by the news coverage, web posting, cards and flowers, and especially the contributions that will secure the future of Cathy and the children.

We will never forget the respect paid to our son during his return to Novato. The sympathy and support from the Novato Fire District, and the people of Novato have been astounding.

Please perpetuate his memory by giving each person you contact respect and kindness.

Thank you,

Patricia and Darrell Rucker

Parents of Steven L. Rucker

veteran firefighter and father of two as they stood in the lobby around a memorial consisting of a photo of their fallen friend and the gear he carried the day he died. "We're all struggling, trying to make sense of the situation," Northern said.

Friends and associates remembered a man who volunteered with the Boy Scouts and organized holiday toy drives.

At a morning briefing at a base camp near Julian, many firefighters wore black bands on their badges in memory of Rucker as they prepared to go battle the blaze that had killed him. Families of Engine 6162 flew to San Diego the next day with Fire Chief Jeff Meston, who said Rucker was "really one of those firefighters that we all love. He was just a great man."

TWELVE

No End in Sight

"WE WERE WARNED . . ."

I t was the subject of numerous *Mountain News* and *Crestline Courier-News* front-page stories, it was the topic of town hall meetings and fire safe council gatherings for nearly two years, and it was the constant source of fear for public safety and emergency officials at every level of government. "It" was the fire still raging out of control in areas and towns all across the San Bernardino Mountains.

We were warned.

At the town hall meetings and at special meetings organized for neighborhood and citizens groups, Fire Marshal Peter Brierty pleaded with mountain residents to become "fire aware" and to do everything in their power to "get your trees down."

Time ran out.

According to Brierty, whom I interviewed by cellular phone while breaking every speed limit across Colorado and New Mexico to get home, the mountain evacuation went exactly according to plan and it couldn't have gone any better. That contrasted with Brierty's own report of the evacuation of Highland, below the mountains, the next day.

"That was more like a circus free-for-all," Brierty recounted. "It wasn't like the mountain residents, who were well organized and proceeded down the mountain just like we'd hoped."

While covering the unveiling of the county Block Grant program in September, before the fire, I had spoken with Brierty about additional fire concerns he had.

He said his next push would be to raise public awareness about shake roofs in the mountains: "Here we are in one of the areas of highest fire danger and just look at the number of shake roofs just in this neighborhood alone."

The fire marshal indicated he knew that would be a tough sell to the public, especially after so many of them had paid so much money for tree removal. This day it seemed like a small price to pay.

And listening to news interviews and word on the street at the evacuation shelters, I realized there was still an inordinate number of the general public who just didn't understand how we arrived at this point. Some would stare blankly, wondering what "they" meant by bark beetle infestation.

As California senator Dianne Feinstein said from the Senate floor the Monday before the fires, "With all these conditions for disaster in place, I fear that California could face a devastating season of wildfires." Sadly, that seemed to be happening as our little rental van screamed down the interstate.

"We need to take action now! Not just to correct our mismanagement of the forest and the brush, but for a more basic reason!" Feinstein exclaimed. "We need to act in advance because of the terrible fact that most of the deaths that will occur in these fires will happen because people will have too little time to escape."

For whatever reason—and now was not the time for finger-pointing—most of us had failed to act in advance.

But we're mountain people, I remembered thinking, *and that means we'll rebuild and restore and things will be good again. They truly will.*

Meanwhile, the flames raged on. There was no end in sight.

CEDAR GLEN IS GONE

By the time the Old Fire made its way down Hook Creek Road and into the tiny Lake Arrowhead area community of Cedar Glen, more than three hundred homes would be destroyed from the intersection of Hook Creek Road and Western all the way to the end of Hook Creek Road near Deep Creek, a distance of about three miles.

The fire would also claim homes near Lakeview Drive.

The flames from the Cedar Glen portion of the fire continued to threaten homes along Emerald Drive and the area near Papoose Lake in Lake Arrowhead and caused extensive damage in the Little Bear Creek area below the dam. Fortunately, the fire moved away from the Mountains Community Hospital.

Flames approached Running Springs from the south side of the mountain just west of Highway 330 and threatened homes located in Live Oak and Enchanted Estates.

While the flames continued toward Running Springs, numerous homes burned in the Valley of Enchantment near Crestline as strong winds blew in a north-northwesterly direction, continuing to threaten Crestline and Lake Gregory.

In Cedarpines Park on the western perimeter of Crestline, numerous homes burned as the fire wrapped around the northern high desert edge of the San Bernardino Mountains and moved toward Lake Silverwood.

Weather predictions called for the winds to be moving in a northerly direction from 25 to 40 miles per hour, with gusts to 50 miles per hour. This was not a good sign. However, good news eventually arrived when a report was issued concerning the possibility of rain or snow in the mountains by Friday night. Firefighters kept their fingers crossed.

Meanwhile, early Sunday morning, October 26, an inferno had engulfed Saturn Way off of Skyland in southeastern Crestline directly on the Rim. More than twenty homes were lost in that area.

By Tuesday morning, October 28, the ground was still hot under the feet of Crestline residents Chuck Forrest and Brad Leonard, two miners who were working on the tunnel project in Waterman Canyon on Saturday morning when the Old Fire broke out.

Tuesday found them walking through an area littered with

burned cars and twisted metal. The only structures left stand-ing in the area were brick chimneys.

Forrest and Leonard first came down to the area from Forrest's house after a friend asked him to check on the house of another friend.

"He said, 'See if her house is OK down there,' and we walked down here to look, and, of course, it's gone," Forrest said.

The longtime Crestline resident lived at Inspiration and Skyland, just above the area of Saturn Way.

"They tried to get us out, but I have five dogs and nowhere to go," Forrest said. He added that the police were insistent in trying to get him to leave. After a while, his girlfriend—who was with him for part of the time—had had enough. "My girlfriend got scared when the cops kept com-ing, so I put a few dogs in her car and she went down to West Covina to her dad's house," Forrest remembered.

He described the scene he saw below his house: "It was bad. . . . It was like a firestorm whirling through here."

Forrest commented from his personal knowledge of the neighborhood as he walked through it. "This guy here, I don't know why he left the truck," he said, indicating the burned-out hulk of what was once a Chevy pickup truck. Forrest pointed from one house to another that was beside it. One had been destroyed and the other was spared.

"This property here was the dad's and the other was the son's," Forrest said. "The dad's is good and the son's is destroyed.

"What I'm doing is just remembering all of these homes," Forrest said. "I've lived up here at Inspiration and Skyland for eighteen years and I know what all of these properties looked like. They were nice old places. It just really breaks my heart."

Forrest and Leonard were also making things right whenever they could. "We went through Horseshoe Bend yesterday and we saw this big cedar tree on fire, so we took somebody's garden hose and put that darned thing out," Forrest said.

Leonard, who lives on Crest Forest Drive near St. Francis Xavier Cabrini Catholic Church, echoed the complaints of several people who stayed behind: that media reports did not reflect the reality on the ground. "The radio reports weren't too bad, but the TV was very, very inaccurate," Leonard believed.

Forrest said that on Sunday morning he and Leonard were walking near the area of Saturn Way. "You couldn't see anything, the smoke was so thick," Forrest said. "You could hear banging, and we didn't know whether it was old trees going up or what. It was like a war zone. A truck just went down and disappeared in the smoke. The smoke was just boiling through here. That's when the sheriff came by and said, 'You're leaving, or we're going to cuff you up and you're leaving.'" Forrest left the area and went to Leonard's home on Crest Forest Drive.

On Tuesday the twenty-eighth, current Crestline Chamber of Commerce president Mike Chilson and his wife, Regina, were still patrolling Crestline as they had been doing for days since the fires began. On Monday, they had a scare when part of the fire threatened Devil's Canyon. "It was actually in the houses above the Hayloft, but it was staying down in the pine needles," Mike Chilson said. "That was yesterday; then on the other side it went up Monument Peak and then it really took off."

The couple had a generator in their house, so they were able to e-mail people down the hill. The Chilsons said that people down the hill kept tabs on their homes through them.

Ace Lake Drive Hardware employee Tom Wilson stood in front of the store on Tuesday morning, drinking a cup of coffee provided by the Stockade, a local watering hole. He was another of the holdouts.

"I was one of the stubborn ones who decided not to go," Wilson said. "I went up to Mountain Storage and I could see the most awesome flame and smoke all the way from where the storage is and it outlined the Rim as far as I could see. It was a disaster, but it was awesome."

Wilson, who lives in Valley of Enchantment, said he had some parameters set for evacuating: "I said to myself that if it got past the ridge above Cedarpines Park, I was going to get out of here, but it never did." Wilson criticized the media coverage for being inaccurate and sensationalized. "What they needed was people on the ground," he said.

The Stockade, a popular hangout on Lake Drive, was going strong as always. The manager there on Tuesday morning, Bill Madsen, said the Stockade went around the clock, serving the holdouts and running a generator to keep the television going with news all of the time. "It was kind of a command central," he recalled.

Mike Chilson recalled that at about 2:00 A.M. that Sunday morning a large contingent of fire apparatus and law enforcement vehicles from all over Southern California began congregating at the bottom of Lake Drive near Lake Gregory.

"At about three o'clock they started asking me for maps, because they didn't have a clue where they were or how to get around," Chilson said.

Chilson walked down to Mountain Pawn, which he owns, got fifty high-quality maps of the area, and handed them out to the fire and law enforcement officers there.

About that time unconfirmed reports that a suspect or suspects had been arrested in connection with the arson start of the Old Fire began circulating. It was important news because by then the fire had been directly attributed to the deaths of four people.

Apparently sheriff's investigators had released a composite sketch Monday of a man suspected of starting the fire, a white male in his early or mid-twenties who was seen driving a light gray van away from the fire's point of origin.

Investigators with the Federal Bureau of Alcohol, Tobacco, Firearms and Explosives also interviewed two men in connection with the Grand Prix Fire but ruled them out as suspects. Within hours, reports would surface that the hoped-for arrests were premature.

The Old Fire had destroyed more than five hundred structures and burned nearly fifty thousand acres by late Wednesday. More homes were threatened as the blaze roared through the San Bernardino Mountains near Crestline, Skyforest, Cedar Glen, Running Springs, and Heaps Peak.

A motorist who stopped at a turnout on Highway 18 reported seeing *two* men in their twenties in the area of Old Waterman Canyon Road, about three-tenths of a mile north of the San Bernardino city limits, about 9:15 A.M. Saturday. He said one of the men threw something in the brush that ignited a fire and then both of the men had climbed into the gray van and fled south on Old Waterman Canyon Road.

Two San Bernardino men—seventy-year-old James W. McDermith and ninety-three-year-old Charles Cunningham—would die from stress related to the fire on Saturday. McDermith collapsed while evacuating his home on Stanton Avenue and died at St. Bernardine Medical Center at 4:24 P.M. Cunningham collapsed while watching his home on Toluca Drive burn to the ground, and was also taken to St. Bernardine Medical Center, where he died at 5:32 P.M.

Also on that fateful Saturday, seventy-year-old Chad Williams collapsed as he and his wife were reportedly scurrying to load items into their vehicle at the couple's Crestline residence. Paramedics from the Crest Forest Fire District received a 911 call and responded to the Williams home, where the elderly man was found to be in full cardiac arrest. He was transported to Mountains Community Hospital in Lake Arrowhead, where he was pronounced dead at 12:25 P.M.

The death of a seventy-five-year-old Big Bear resident, Gene Knowles, would become the fourth known fire-related death in San Bernardino County when he died at 3:15 A.M. Sunday, October 26.

The San Bernardino County Sheriff's Department continues to lead the arson investigation, which is now also considered a murder investigation because of the four fire-related deaths.

The Sheriff's Department spokesperson, Chip Patterson, explained, "Because of the four deaths to date, whoever is arrested in connection with the arson-caused 'Old Fire' will face felony murder charges."

As of this writing, the case officially remains open; however, investigators believe they may already have the man responsible for starting the Old Fire in custody.

Raymond Lee Oyler, a thirty-six-year-old Beaumont, California, auto mechanic, became a person of interest in the Old Fire case when he was arrested and charged with intentionally starting the Esperanza Fire in Riverside County.

In the Esperanza blaze, which killed five firefighters, Oyler was charged with five counts of first-degree murder and seventeen counts of possession of materials to commit arson. In addition, the Riverside County district attorney's office charged Oyler with twenty-three arson related felonies connected with the Esperanza Fire and ten counts of arson for a series of smaller fire starts in the months preceding the larger blaze. It would take days to extinguish the Esperanza Fire due to the fierce Santa Ana winds coming off the desert, and in the end the massive blaze would consume forty thousand acres; an area exceeding sixty square miles.

According to published accounts, the San Bernardino County sheriff's homicide investigator in charge of the Old Fire case told an area newspaper that Oyler first became a suspect when the Riverside County district attorneys drew the detectives' attention to physical similarities between Oyler and the composite sketch that had been circulated within hours of the start of the Old Fire. Oyler's attorney, Mark Raymond McDonald, who emphatically stated to the media that his client was not responsible for starting any of the fires, denied any physical similarity or resemblance between his client and the composite sketch.

Nevertheless, whether Raymond Lee Oyler is ever charged

with setting the Old Fire, the case against him is moving forward in the Esperanza Fire. The likelihood that Oyler may ultimately face the death penalty increases as prosecutors continue to amass evidence that the former Beaumont auto mechanic was the person responsible for a string of arson fires.

But whatever happens to Oyler, it will not return firefighters Pablo Cerda, Daniel Hoover-Najera, Mark Loutzenhiser, Jason McKay, or Jess McLean to the love and safety of their families.

On Monday, October 27, Senator Feinstein joined Governor Gray Davis and San Bernardino County Board of Supervisors Chairman Dennis Hansberger in requesting a Federal Emergency Declaration from President George Bush due to the devastating effects of the fires. The President came through.

Federal aid was on the way, with details to be administered through the Federal Emergency Management Agency.

The government would inherit a handful.

In San Bernardino County alone, what had started as a 280-acre fire deep in Waterman Canyon quickly ballooned into one of the worst fires in California history, growing to over twenty-eight thousand acres and engulfing and destroying nearly a thousand homes in the San Bernardino foothills and Crestline and Lake Arrowhead areas. And the Old Fire wouldn't even claim the biggest prize. That dubious distinction would go to the Cedar Fire down in San Diego County.

Our own story of recovery would prove as byzantine in nature as the thousands upon thousands of other personal stories circulating in the weeks, months, and now years following the fires.

My family and I returned to the mountain on November 9, 2003. We were very fortunate, as we had a place to go.

In April of 2003 I was assigned to do a story about a tree cutting company that had just signed an advertising contract with the *Mountain News*.

I met Jeff and Patricia Gruett at the Lake Arrowhead home they had recently purchased as a "bunkhouse" for their logging crew. For the next few months Gruett Tree Service would be engaged in cutting down bark beetle-infested trees all over the mountain. Little did I know the home that served as the interview location for the article would become our home for the next two years.

When the Gruetts learned we had lost our home in the fire, they packed up the crew and offered us unlimited use of the house. We could keep it as long as we wished, and the Gruetts only charged us rent equal to their mortgage payment. To put it in perspective, the average rent on a house in Lake Arrowhead is between $1,800 and $2,500. The Gruetts charged us just $975 per month.

Immediately offers began pouring in for free furniture, clothing, toys for Gracie, and small appliances.

An organization was created called Rebuilding Mountain Hearts and Lives, which is still going strong and has since become a community clearinghouse of information, something that was needed right after the fires and is sorely needed now as the county embarks on Cedar Glen redevelopment plans.

The organization, headed by David Stuart and Ira Maser, has become an ombudsman of sorts, helping fire survivors interface with the county at this critical time.

We moved into our new home, still shocked by the fire, but we were functional because of the outpouring of great caring and love that was directed at our family from people who before the fires were merely casual acquaintances or even strangers. Now they were friends, and they remain so to this day.

The government wheels began to grind, and even Uncle Sam would step up to the plate and help our family and the families of our friends and neighbors.

Within a week we had a rent assistance check, and by New Year's Day they would follow that up with a lump payment to help us get back on our feet. In our eyes, America was at her best in the weeks and months following the fires. There were some bumps in the road, but we made it.

Two years later, when Hurricane Katrina would blast her path across the Gulf Coast, we wondered why the government agencies that had helped us so much during our disaster/crisis were unable to adequately assist our New Orleans brethren.

THIRTEEN

A Turn in the Weather

A break in the weather on Friday, November 2, finally gave the firefighters a crucial edge, leading to victory over the hellish blazes that had ravaged Southern California for nearly two weeks.

Because of a dense fog from the coast that blanketed most of the fire areas, lower temperatures, and tamed winds, firefighters could finally take the time to bulldoze some buffer zones.

The weather forecasters said heat and dry desert winds that helped create the deadly infernos could possibly return. Fortunately, the fire lay down.

"We've got a sleeping giant out there," Forest Service spokeswoman Sue Exline said, but her caution gave way to glee when six inches of snow fell in the mountains the next evening. Winds gusting to thirty miles per hour were predicted, but they stayed away too. The incident commanders pulled most firefighters off the line for the night on Friday as temperatures in the mountains plummeted into the low twenties.

Big Bear remained in some danger as the remnants of the Old Fire moved to within six to eight miles of the main business district. Some fifteen thousand people had been evacuated and had yet to return to the resort town. "The fire is just creeping around, not making those big runs that we had seen," Exline said.

Fire crews took advantage of the lull and cut thirty miles of firebreaks to protect the small communities surrounding the lake and removed brush and trees from zones one hundred feet wide to deprive the fire of fuel. The fire lines had been as close as fifty yards to some of the homes.

"This is an opportunity," Exline said. "We can get in there in the next forty-eight hours to fight the fire on our own terms, without the forces of the weather."

In spite of the good turn for San Bernardino Mountains residents, there were seven fires still burning across the four counties. Images taken from satellites showed the winds had carried the smoke as far north and east as the Plains and the Great Lakes.

By now the largest blaze—the 275,000-acre Cedar Fire— was 65 percent contained. Firefighters were elated that the threat had at least eased against Julian, the old Gold Rush tourist town with the gorgeous apple orchards.

As the danger seemed to ease in San Diego County, people who had fled their San Bernardino Mountains homes began clamoring to return amid reports of looting and trespassing. Three women held signs reading: "Stop the looting. Let us back in our homes," hoping to convince the local authorities to let them return to Crestline. San Bernardino County sheriff's spokesman Chip Patterson denied that looting was a problem in the mountain communities affected by the fire.

While thousands had fled the Big Bear area, a few people had stayed behind to protect their property.

Kelly Bragdon remembered sitting at the bar at the Log Cabin Restaurant, sipping a beer and watching news reports of flames blazing through the forest.

"I've got too much to lose to leave here," Bragdon said. "I don't think we're jeopardizing anybody's lives but our own, just trying to save what we've got, everything we've worked for."

Craig Brewster, owner of the Robin Hood Resort hotel just off Big Bear Lake, stayed in town and opened up his place for firefighters.

"It was a ghost town, strange," he remembered. "It was a weird feeling. We'd typically be getting ready for trick-or-treaters."

As that awful October of 2003 gave way to November and thoughts of Thanksgiving, Ellen Bechtol looked to the dark clouds over the mountains and hoped the cooler weather would finally tame the raging fire. Bechtol and her family had been evacuated from Running Springs on Monday the twenty-seventh and lived at the evacuation center at the San Bernardino Airport, waiting for word about when they could return.

She also remembered how the wildfires brought together outgoing governor Gray Davis and the new governor-elect,

Arnold Schwarzenegger, when they toured the disaster relief center that had become her temporary home.

"For the next two weeks before Mr. Schwarzenegger takes over we are going to do everything possible to put victims back on their feet," Davis said. By the day of the visit about fifty-four hundred people had already applied for assistance and more than one hundred thousand dollars had been doled out to the fire victims. "Clearly, a lot more money is going to go out the door," Davis promised.

Schwarzenegger recounted a tour he had taken just days before of the San Bernardino Mountains, where he met with about four hundred firefighters before they returned to work.

"It was a very special moment for me because those firefighters are true heroes," he said.

But some recall being less than impressed with the visit.

"It was just a diversion," recalled Merlin Duvall, who fled his home of seven years in the Claremont neighborhood of Palmer Canyon with only two blankets, a pillow, and some documents. "But I guess it gave me something to think about other than what I lost."

Duvall remembered it took him several hours to navigate his way through the different agencies set up to help people get back on their feet. He said he talked to the Lions Club, the Red Cross, the Small Business Administration, and the Federal Emergency Management Agency. "All they can do is deal with immediate housing and money," he said. "As far as getting your life back, that was a personal thing."

In the waning days of the fires, one assignment continued to pick up steam—the directive to find the Old Fire arsonist. There were more than five hundred tips, dozens of cars

THE MOUNTAIN FIRES

What follows is a brief history and day-by-day recap of the fires in the San Bernardino Mountains in 2003:

TUESDAY, OCTOBER 21
The Grand Prix Fire was reported near Grand Prix Drive and Shetland Lane in Northern Fontana in the community of Hunter's Ridge at 2:22 P.M.

The first units on scene reported two acres on fire and a rapid rate of spread fanned by a strong southwest wind. The fire grew to 100 acres in the first hour and was 825 acres by the end of the day.

WEDNESDAY, OCTOBER 22
The Grand Prix was now reported to be 1,958 acres. Over seventeen hundred structures were threatened. Lytle Creek Road was closed.

THURSDAY, OCTOBER 23
The Grand Prix Fire's burned acreage report stood at 3,500. The tiny town of Lytle Creek was evacuated. Two major power lines were lost.

FRIDAY, OCTOBER 24
The Grand Prix burned acreage was now 12,600 and growing. Interstate 15 and I-210 were closed. Parts of

(continued)

THE MOUNTAIN FIRES (continued)

Rancho Cucamonga were evacuated. Another high-voltage line was lost.

SATURDAY, OCTOBER 25

The Grand Prix burned acreage grew to 27,180. San Antonio Heights along with western Rancho Cucamonga were evacuated. Five homes were destroyed.

At 9:16 A.M. (approximately) the Old Fire was reported in Waterman Canyon, and it rapidly spread downhill toward the Arrowhead Springs Resort. The fire quickly moved toward the community of Del Rosa in northern San Bernardino.

Santa Ana winds pushed the fire east and west of Highway 18 along the foothills. The forest supervisor ordered the San Bernardino National Forest closed to all use. Evacuations took place in San Bernardino and the mountain communities from the Crestline area to Lake Arrowhead. The fire was at 10,000 acres by 6:00 P.M. and hundreds of homes were destroyed.

SUNDAY, OCTOBER 26

The Grand Prix was at 52,184 acres, and the number of homes destroyed rose to sixty. In addition to the I-15 and I-210 freeways, the I-215 as well as railroad lines in Cajon Pass, which handles more railroad tonnage than any other railroad pass in the world, was closed. The fire moved toward Claremont and Mount Baldy Village in the San Gabriel Mountains. A decision was made to call the portion of the fire in Los Angeles County the Padua Fire.

The Old Fire stood at 24,000 acres, and backfiring was initiated along Highways 18 and 138 near Crestline. More than three hundred homes were lost in the fire at this point, and evacuations continued.

More than twenty-two evacuation centers were in use throughout Southern California. The Grand Prix Fire was officially merged with the Old Fire.

MONDAY, OCTOBER 27

The Grand Prix Fire was at 57,332 acres. The Union Pacific and the Burlington Northern and Santa Fe railroads were still shut down in Cajon Pass. The communities of Lytle Creek, Rancho Cucamonga, San Antonio Heights, Upland, Mount Baldy, Claremont, Rialto, Fontana, La Verne, and San Dimas continued to be threatened.

The Old Fire was at 26,000 acres, and structures continued to be burned as the fire spread rapidly throughout the area. A secondary evacuation center opened at Sultana High School in Hesperia, on the northern side of the mountains in the high desert. By the end of the day, more than sixteen hundred firefighters were assigned to the Old Fire.

(continued)

THE MOUNTAIN FIRES (continued)

TUESDAY, OCTOBER 28

The Grand Prix Fire was at 59,229 acres and evacuation restrictions were lifted for the southern areas of the fire, but Lytle Creek remained under an evacuation order.

The Old Fire grew to 36,780 acres, and the fire flame lengths were reported as being between thirty and fifty feet, with "spotting" from five hundred to one thousand feet ahead of the main fire. The fire had now burned into the Lake Arrowhead area and Running Springs.

WEDNESDAY, OCTOBER 29

The Grand Prix Fire was at 59,229 acres on this date, and evacuations in Lytle Creek were lifted.

The Old Fire was at 47,960 acres, and by now active crown fires and flame lengths in excess of two hundred feet were observed in the Silverwood Lake and Miller Canyon areas on the north side of the San Bernardino Mountains. Evacuations in the northeast Del Rosa, East Highlands Ranch, Yucaipa, and Muscoy areas on the south side of the San Bernardino Mountains were lifted; however, mandatory evacuations of Oak Hills, Baldy Mesa, and Summit Valley on the north side were issued.

Voluntary evacuations were ordered in the southern portion of Hesperia. Additional evacuation centers were opened at Apple Valley High School and the Hesperia Fairgrounds. By the end of the day an additional 350 homes were lost, 314 of them in the Cedar Glen area.

THURSDAY, OCTOBER 30

The Grand Prix Fire grew slightly to 59,358 acres, the Old Fire to 91,281 acres. The fires continued moving unimpeded, exhibiting rapid uphill runs and spotting.

More than forty thousand residents remained evacuated.

Two more federal teams were ordered for contingency planning and preparation in the Big Bear area.

The weather finally turned, with lowering temperatures and increasing humidity, allowing crews to start construction of containment lines in the north Lake Arrowhead area. Evacuation orders were lifted in the Baldy Mesa and Oak Hills areas.

FRIDAY, OCTOBER 31

The Grand Prix Fire remained at 59,358 acres, and the Old Fire remained at 91,281 acres as the weather continued to give firefighters a break, allowing the crews to aggressively attack the fire. Dozer lines were completed in the Big Bear area. Arson was established as the cause of the fire.

SATURDAY THROUGH TUESDAY, NOVEMBER 1–4

The Grand Prix Fire rose to 59,448 acres and was reported as 98 percent contained. All evacuation orders were lifted on November 4, 2003.

The Old Fire remained at 91,281 acres; however, it continued to threaten the Lake Arrowhead and Running Springs communities. Mountain area residents began to return to their communities.

stopped, a handful of men questioned, at least one confession, and endless rumors.

The San Bernardino County Sheriff's Office did manage a description of a suspect, whom witnesses would describe stepping out of a gray van and dropping something into the brush causing a fire.

He would then step back into the van before speeding away.

Dozens of federal, state, and local arson detectives worked around the clock. They called for the public's help and talked to witnesses at the evacuation centers. The investigators immediately began to study burn patterns where the fires began, and still the arsonist managed to elude capture during the critical first hours and days of the investigation.

Forest Service agent Jerry Moore said a man had confessed to starting a Ventura County fire that burned three homes and sixty-eight thousand acres, but the case still remains under investigation. "Anybody can come in and say, 'I did it,'" Moore said. One could ponder the reasons why, but the arsonist is a strange creature, to be sure, and investigators could find no obvious motive for the arsons. Catching wildfire arsonists can be difficult because they can set time-delay devices or wait until the coast is clear. Usually when investigators get a good lead, Moore said, it's because a witness just happened to be driving by.

San Diego County authorities, meanwhile, issued a formal statement that the Cedar Fire, which killed sixteen people, including firefighter Steven Rucker, and burned nearly fifteen hundred homes, was sparked by the lost deer hunter, Sergio Martinez, who admitted he had fired a signal gun.

FOURTEEN

Without a Chance

THURSDAY, OCTOBER 30, 2003

Residents at the end of Muth Valley Road in the community of Lakeside—a gorgeous little town about twenty-five miles east of San Diego—found they were in a hellish landscape of embers and flames when suddenly rousted from their beds in the middle of the night. What they would each decide to do in the coming moments would literally determine life or death.

Some would make a determined run for a nearby reservoir. Others jumped into a swimming pool or tried to speed away on the only escape route. One family chose to wait out the fire parked in their driveway. Four died, the most in any single location in what was until then the most destructive wildfire in California history.

The small community of Lake View Hills Estates, made up of only ten homes, was one of the first places overrun. By morning it would be a monument of gray and black landscape, blasted trees, and bare shrubs stripped by flames. The fire came roaring up over the hill from Wildcat Canyon shortly after 3:00 A.M.

The huge San Vicente Reservoir below the quiet hillside community seemed a beacon of salvation. It was a quarter mile away down a trail lined with thick brush. The Shohara family, James, Solange, and their son, Randy, tried to make their way there.

Bob Daly thought about joining them but changed his mind. Daly, seventy-five, and his wife instead decided to take their chances and dived into the pool. The remaining neighbors headed down Muth Valley Road, which was the only way in or out.

Standing at the community's security gate, retired firefighter Larry Redden saw fire consuming the road ahead and screamed at his neighbors to turn around and head in the other direction. But Stephen Shacklett couldn't hear his friend. Last seen behind the wheel of his motor home, the fifty-four-year-old, accompanied by his two wolfhounds, was already rumbling down the road into the heart of the fire.

Stephen Hamilton, forty-three, and the others turned back. Hamilton's wife, Jodi, who was six months pregnant, thought the reservoir would be their haven. "We've got to get to the lake!" Jodi screamed.

But Stephen, a vice president of a construction company,

knew the path could prove dangerous. The couple and their two-year-old son rode out the fire parked in the driveway in their sport-utility vehicle.

"It was a tornado of fire," Stephen Hamilton recalled, "a swirling, scorching tornado."

The firestorm at its apex lasted for about half an hour, he said, and then moved on, the wall of fire headed toward San Diego. The best way out, it seemed, was to stay put. Daly and his wife stayed in their pool and then climbed out once the flames passed. But the Shoharas never made it to the reservoir. "They were incinerated to just skeletons," Stephen Hamilton recalled.

What was left of Stephen Shacklett's motor home was about a mile down the road, surrounded by yellow crime scene tape. The steel frame was twisted and melted to the ground. "He couldn't back up. He couldn't go forward," said his son, Steve Shacklett Jr., as he surveyed the wreckage of his father's home.

"You just wonder what goes through a guy's mind," the older Shacklett's brother, Cliff, said as he began to choke with emotion. "Probably all that was on his mind was his son." Stephen Shacklett's girlfriend, Cheryl Jennie, remained at the house. "She had two ways to go: One was the good way; one was the bad way," Cliff recalled. She took the good way.

ENGINE 31

During a rare lull in the storm, the Engine 31 crew took advantage after a hard week spent nonstop on the fire line.

Awaiting orders while stopped along a highway shoulder, Crew 31 traded stories, enjoyed a little nourishment, and spent some time rehydrating. Someone brought a case of drinking water to the line, and they took advantage of it. The crew even caught a few winks of precious shut-eye. Recess didn't last long. A new assignment order was issued, and they were told to "head for the mountain community of Julian and save what you can."

They soon found themselves standing before the beast, fighting to get a leg up on the greedy flames. They had one thing on their minds: Save lives and property. The one woman and four men of USFS Engine 31 were exhausted and frustrated but were not going to consider defeat as an option even for a moment.

When the crew was initially dispatched from their home base at the Angeles National Forest, it was to help fight one of many fires dogging Camp Pendleton, situated between San Diego, Orange, and Riverside counties. But that was before even more deadly conflagrations had surfaced across Southern California. They had been putting in sixteen-hour days helping to save hundreds of homes and structures.

"It was more than I expected. I thought we'd have a half hour to work on a house. We had about five minutes," Crewman Sanchez, age twenty-five, would say. "We just didn't have any time because of the winds." Sanchez, even at his young age, was considered a veteran in the crew, having fought fires for nearly seven years. "This is some of the most stressful firefighting I've done." Having hopped from house to house in one San Diego mountain community, the crew made a huge effort to remove debris and down the trees that acted as matchsticks in a fire of this magnitude. But they were repeatedly frustrated by fire that would roar past as soon as they left one house to start work on another.

They would soon help with a burn-out operation intended to deny the fire additional fuel. They would do the best they could before heading off eastward to Julian.

Their Forest Service rig would join a caravan of engines from all over the state. Following a winding gravel road just south of the historic gold-mining town, Engine 31 pulled into the driveway of a new home. Its owner had draped a water hose on the roof, which only dripped water onto the back porch. The flames moved toward the house from the south, and a neighbor's woodpile was already burning next door.

Following a quick assessment of the property—no visible fuels that would put the home at risk, a good defensible space had been created—Engine Captain Tracy McGuff made a decision, like a doctor performing triage, to take her crew somewhere else. "Guys, this house is looking good," she said. They headed one block south to a home on Glenside Road.

A wood lattice along the side of the house was quickly tossed, along with clinging vines along the wall. As one crewman took his chain saw to the box shrubs, Sanchez and another firefighter began moving items on a balcony they feared might catch fire.

Using his pocketknife to cut down a row of sunshades draped around the second-story patio, Sanchez tossed them away from the house.

A loud sound, akin to a sonic boom, was heard to the south. A geyser of bright red flames exploded into the sky.

"Propane tank," Sanchez recalled. The house was now secure enough so the crew could address the nearby brush. Sanchez started unraveling hose from the engine.

"It's the fastest-moving fire I've ever seen. It's huge," said crew member Steve Pera, age twenty-two. "We just let the horses out at one place and said, 'Good luck, guys.'" After creating a line in the brush, the crew felt confident enough to declare this house as one that would make it.

McGuff rounded up her squad and they mounted their rig. Engine 31 was off again, headed for another hellish foray into the fire.

SOME OF THE DEAD were discovered in their burned-out cars or just outside them. Some were found lying next to the pets they had hoped to save. Some had collapsed from heart attacks brought on by the stress of evacuation or watching their homes burn.

"This fire was so fast," Glenn Wagner, San Diego County's chief medical examiner, remembered. "I'm sure we're going to find many folks who simply never had a chance to get out of their houses."

Ashleigh Roach, sixteen, died Sunday, October 26, while she and her brother tried to drive away from the Paradise Fire, which destroyed her family's home in Valley Center, a rural community in inland northern San Diego County.

The high school junior's brother, Jason, was driving through the smoke when their car hit another car and then a tree. Jason, twenty-two, escaped with minor burns; Ashleigh was trapped inside. Their sister, twenty-year-old Allyson, was seriously burned while trying to save a cat. Her friend, a twenty-one-year-old Marine named Steven Lovett, had to push her into his car to get her away. He was treated for third-degree burns.

At the San Diego County morgue, a team of investigators worked with dental records, old X-rays, jewelry, and even licenses found on nearby pets in a seemingly endless effort to confirm the identities of the people who became trapped in flames. "They were all carbonized, which means that they were well on their way to being cremated remains," Wagner said.

Nancy Morphew, fifty-one, was a Valley Center horse rancher who died as she attempted to drive away from her home.

Galen Blacklidge, fifty, of Moreno, died while trying to escape in her vehicle.

Mary Peace died on the Barona Indian Reservation just north of Lakeside.

Most of the confirmed deaths involved people who were consumed by flames in or near their cars, Wagner said. "It's clear the fire overcame them," he said. "They got out of the vehicle and tried to flee, and were knocked down by the flames, basically where they stood, oftentimes two or three feet away from the vehicle."

At least five victims were found with the remains of their dogs at their side. Wagner said he kept the animal remains with their owners. "Families usually feel very strongly about their pets, so I treated them the same way," he said. "They lived together and they died together."

. . .

AS FLAMES CLOSED IN, a neighbor ran to Nancy Morphew's door to warn her. But Morphew didn't seem worried about getting out in time.

"She met me at the front door and said, 'I know, I know. Go help other people,'" David Wallace recalled. "She seemed like she had a plan."

After the fire tore through, Morphew was found dead in the road, her burned-out truck sitting at the bottom of a steep ditch.

Like many of the people who died in the wildfires she took pride in running her own horse ranch, and apparently underestimated how fast the flames were moving.

VICTIMS REMEMBERED

From the Lakeside Historical Society Web site: In the Cedar Fire, our community lost its sense of security as many homes burned to the ground. But worse, we lost the lives of so many friends, neighbors, and loved ones.

- Galen Blacklidge, 50, Moreno, teacher
- Christy-Anne Seiler-Davis, 42, Alpine
- Gary Edward Downs, 50, Lakeside, small-business owner
- John Leonard Pack, 28, Lakeside
- Quynh Yen Chau Pack, 28, Lakeside
- Mary Lynne Peace, 54, Lakeside, nurse

"I think she got disoriented and the smoke was probably real bad," recalled another neighbor, Charlene Pierce. "It looks like she just drove into the canyon."

Morphew was apparently overcome by flames after making her way out of the ditch, according to Medical Examiner Wagner. Her empty horse trailer was spilled about twenty-five yards behind the truck. Neighbors believe she tried to escape before loading the horses.

Others who lived on her street recalled how they were caught off guard by the flames, which roared up from Hell Hole Canyon with little warning.

Gary Olson had been watering down his house when the

- Ashleigh Roach, 16, died in Valley Center, student
- Steven Rucker, 38, Novato, firefighter, died in Julian
- Stephen Shacklett, 54, Lake View Hills Estates, construction superintendent
- James Shohara, 63, Lakeside, correctional officer
- Randy Shohara, 32, Lakeside
- Solange Shohara, 43, Lakeside, correctional officer
- Jennifer Sloan, 17, Lakeside, student
- Robin Sloan, 45, Lakeside, Wal-Mart employee
- Ralph Marshall Westley, 77, Lakeside, retired retail clerk
- Unknown migrant worker found mid-December in the I-15 SR-52 area

fire arrived. "You just felt a gust of hot wind. It was one big flame. It was moving so fast you didn't have time to think," he said.

At Ashleigh and Jason Roach's house, a barn was on fire, sending flames across their driveway, neighbors recalled. Then the winds blew flames across the road, forcing residents to drive through the fire. "It was like a dragon just huffing and puffing," recalled Dan Contreras, a forty-one-year-old plumber. "It wouldn't stop. It just kept coming."

FIFTEEN

The Fires Wind Down

In two decades of fighting fires from the sky, air tanker pilot Peter Bell had never seen anything like the vortex in the Southern California skies. "There was a big spiral, like a tornado, that sucked all this dirt and garbage into the sky," he recalled.

The debris cracked windshields on six tankers, and cockpits filled with smoke. Another pilot saw a four-by-eight-foot sheet of plywood sail past at fifteen hundred feet.

Thirty-five air tankers and eighty-six helicopters fought the flames of the Southern California wildland fires. Their key role was to dump their three-thousand-dollar loads of retardant on the outskirts of the fires to help crews build fire lines around the flames.

To many of the pilots, the fires that ravaged Southern California were among the most intense they had ever flown through. Air tanker pilots are usually employed to douse small fires before they spread. Here pilots were forced to fly through narrow canyons, thick smoke, and high winds. It was so bad at one point that due to the amount of windswept debris they encountered, the pilots requested reconnaissance planes to fly ahead of them on missions.

"People think we're daredevils, but we're not," said Bell, a pilot from Missoula, Montana, who worked under contract for the U.S. Forest Service. "All we do is practice safety, safety, and safety."

The pilots who flew the missions said they were stunned by the damage caused by the blazes. "It looks like the surface of the moon. There's not a stick of wood," said tanker pilot Jim Cook. "Have you ever seen pictures of Hiroshima? That's what it looked like."

PROSECUTORS ASKED A FEDERAL judge to reject defense efforts to move the pending trial of Sergio Martinez, the lost hunter accused of starting the Cedar Fire, to another venue, away from San Diego.

Lawyers for Martinez asked for the change of venue, arguing that "hostility, hatred and ill will" stirred up by the media coverage would make a fair trial impossible in San Diego. The U.S. Attorney's Office contended in court papers made public that it was not necessary to move the trial because although almost everyone in San Diego was aware of the fire, "few are prejudiced against the West Covina man."

"The publicity in this district, while extensive when it comes to the Cedar Fire itself, has barely addressed the culpability of Martinez," prosecutors said, citing a review of 871 articles about the fire in *The San Diego Union-Tribune*. Of those, 31 mentioned Martinez.

"The reporting has been largely factual in nature, has not involved a call for a conviction and has not involved the disclosure of inadmissible evidence," prosecutors Michael Lasater and Kevin Mulcahy wrote. But defense lawyer Wayne P. Higgins of Los Angeles said saturation media coverage of the fire and Martinez make his job untenable. "Jurors will be unable to separate their own personal feelings of hatred and antagonism toward the defendant from their judgments of the facts," he wrote.

In the end, none of the objections from either side would matter because a jury would never be impaneled to hear the case.

In March 2005, nearly two years after the hellish days of October 2003, Sergio Martinez, in exchange for a plea of guilty on two charges: making false statements to a law enforcement authority, and the willful setting of a fire in a national forest, decided to throw himself on the mercy of the court.

It appeared mercy followed Martinez to court on Thursday, November 17, 2005, his formal sentencing date, when United States Attorney Carol C. Lam announced to the media and family members assembled at the courthouse that United States District Court Judge Roger T. Benitez had rendered his decision: Sergio Martinez would receive a sentence of five years probation, six months in a halfway house, a fine of nine hundred dollars, and nine hundred hours community service.

Many were outraged, including Cedar Fire victim Kelly Williams who had lost everything she owned, including her home, when the blaze swept through Wildcat Canyon. Williams felt Martinez should have been ordered to spend five years behind bars. "If you spend five years in jail, it's like spending one day for each house you burned down," Williams said, as quoted in numerous media accounts of the sentencing hearing.

Williams did acknowledge, according to Associated Press and television reports released later in the day, that she believed Martinez when he had earlier addressed the court and the surviving victims offering what she described as a heartfelt apology. "I was so glad when you said [you were sorry]," Williams said. "When you said it, I could hear in your words that you meant it."

Williams went on to say that she forgave Martinez, but still believed he should serve some lengthy prison time.

Then Judge Benitez addressed the courtroom: "If I thought Mr. Martinez set the fire maliciously, five years would not have been enough.

"The time that I impose, if any, is not going to bring any loved ones back, it's not going to bring any property back . . . and I was struck by the fact that Mr. Martinez had taken a course in hunting safety in 2002 and had learned from the course to set a fire if he got lost. . . . This is different than a negligent defendant who tosses a cigarette butt out the window."

Benitez addressed a number of fire victims who had

requested the judge issue the maximum five-year prison sentence. The victims felt a message needed to be sent to deter other careless individuals.

"I'd like to send a message," Benitez began, "that the law doesn't lack compassion and understanding.

"I've struggled with the case for several months now, and wondered if it would be just to send a man to prison for five years, and see him lose his longtime job." And then Benitez announced his decision. Martinez, outside of the courthouse, cried in relief that "God has given me a second chance."

Moments later Martinez and his family were forced to return to the safety of the courthouse. An agitated stranger, apparently upset by what he believed was a lenient sentence imposed by Judge Benitez, cursed and yelled at Martinez, following him all the way back to the building. The judge's ruling elated Martinez, but he had to know at that moment that there remained a long road ahead before the public's perception of him would change.

THE CEDAR FIRE SURVIVORS in Poway were recognized with commemorative plaques to hang on the walls of their homes. Poway lost fifty-four houses to the blaze when it raged across the city's southeastern corner. The enormity of the disaster brought the city together in an unprecedented relief effort that helped those who lost homes and belongings in the fire. Today the city is well on its way to recovery from the blaze. City officials, however, decided they wanted to do something to honor the Cedar Fire survivors.

The plaques are similar to those hung on the outside of

historic homes. Each of the six-inch-by-six-inch bronze-colored squares is engraved with a message that identifies the home in question as having risen from the ashes of the Cedar Fire. "From the end springs new beginnings," the phrase reads.

THE RESIDENTS SUING THE county and the state for upwards of $100 million in damages caused by the deadly Cedar Fire revised their civil suit to narrow their claims. In the newest claim, filed at the East County courthouse, the plaintiffs cite negligence as their reason for suing San Diego County and the CDF.

In the amended version of the suit, the plaintiffs claim that both the county and the state failed to dispatch emergency workers quickly enough and that 911 operators gave callers false and misleading assurances that help was on the way in the early stages of the wildfire.

The plaintiffs also claim the government bodies did not properly control and manage the brush and vegetation within the Cleveland National Forest prior to the fire.

The massive Cedar Fire, sparked on the federal forest land on October 25, 2003, left at least sixteen people dead and destroyed more than twenty-two hundred homes as it sprinted across 275,000 acres in parts of northern and eastern San Diego County.

As they had in the first version of the suit, the fifteen plaintiffs in the civil suit are also asking the court to give the suit class-action status to include anyone who suffered losses in the blaze.

The plaintiffs' lead attorney, Chicago-based Mark Grotefeld, says that the amended claim includes more detail to back up their claims of negligence. "It clearly shows how the county and the state permitted a small, manageable fire to spread when it had the time, opportunity, and weather conditions to have stopped it in its tracks," Grotefeld said.

THE PARADISE FIRE BLACKENED about fifty-six thousand acres (eighty-eight square miles), and destroyed 117 homes and 116 outbuildings. The strong winds pushed the blaze east, but its advance was stopped at state Route 76 at the base of Palomar Mountain in northern San Diego County.

The fire forced the Palomar Mountain community and La Jolla Indian Reservation to evacuate to the Red Cross shelters at Valley Center High School and Emmanuel Faith Chapel in Escondido.

Fire officials at the nearby command center asked for CDF air support near Lake Henshaw but were denied due to dangerous conditions.

Resolution on the Paradise Fire still eludes authorities.

The Paradise Fire was the second largest of the three blazes that swept across San Diego County during the October firestorms. It started the day after the larger Cedar Fire, in and around rural Valley Center. The fires tore through Indian reservations in the Valley Center area and threatened Escondido.

Authorities believe the Paradise Fire likely resulted from carelessness rather than a deliberate attempt to start a wildfire. No suspects have been officially identified.

. . .

THE FAMILY OF ASHLEIGH ROACH filed a claim against the Rincon Tribal Council and a lawsuit against the operators of Harrah's Rincon Casino in connection with the Paradise Fire. The family claims the refusal by the tribe and casino to provide fire protection paved the way for the spread of the fire that killed their daughter and severely injured their son, Jason.

A team of seven investigators has interviewed thirty witnesses, none of whom were actual eyewitnesses. Some saw the flames moments after they erupted at 1:30 A.M. on the Rincon Indian Reservation, a half mile south of Harrah's Rincon Casino & Resort.

The fire originated in a grass- and brush-filled area that sits on rugged terrain in an area known as Snake Mountain. The area is accessible only by a tiny unnamed dirt road off Valley Center Road, near Paradise Creek Road.

Michelle Hall, a neighbor of the Roach family, said she was lucky to have escaped the oncoming flames with her husband and three children after deputies told them to evacuate at 8:30 the morning of the fire.

Three months later, Hall, who lives two houses away from where Ashleigh lived, said it was still difficult to talk about the blaze.

SIXTEEN

Hell Continued

More than two years after the firestorms in Southern California, another global warning of humankind's vulnerability came—this time in the Texas Panhandle.

The devastating fires that hit Texas in March 2006 had been brewing for more than a year. Historically, March storms in the northern finger of Texas have been known to bring pounding rain, hail, and late-season ice and snow.

But global warming brought a new kind of storm to northern Texas in 2006. Borne on hot, dry winds, the biggest firestorms in Texas history brought terror, destruction, and death across nearly one thousand square miles of the Panhandle.

For days, the fires were unstoppable and firefighters were nearly powerless as winds of up to 50 miles per hour sent flames racing through ranches, grass, and brush, sending people fleeing. Fire crews had enough trouble just physically staying ahead of the flames, and seven firefighters were injured in their efforts.

The charred remains of four oil workers found a few days after a fire raged through a Roberts County ranch graphically illustrated the ferocity and speed of the fire that had swept through. The four had apparently been driving down a gravel road when their car was overcome by smoke. Panicking, they ditched the car in a ravine, jumped out, and then began running for their lives. All of the bodies were found within fifty yards of the car.

As an area nearly the size of Rhode Island was blackened, eleven people lost their lives, and dozens of homes and other buildings were destroyed. Throughout the area, the carcasses of cattle littered fields. Some were found entangled in fences that had ended their flight as they desperately tried to race the flames.

The losses, while significant, were limited because this outbreak of fires occurred in sparsely populated ranch land in the north of the state. The Texas Panhandle at the time of the outbreak was considered to be in a mild to moderate drought. The heavily populated Gulf Coast of Texas, however, was in a state of severe drought at the time, and officials there were wary of fire. During the outbreak, fire officials elsewhere in the state were also worried about the drought and tinderbox conditions in the urban areas around the capital of Austin in the central part of the state.

As he toured the Panhandle fires, Texas Forest Service Director James B. Hull was quoted in the *Austin American-Statesman* warning that if a similar outbreak struck forested and urban Austin and Travis counties in central Texas, the result could be devastating.[1]

"Austin is going to be the worst catastrophe Texas has ever seen," Hull said. "With the conditions we're having in the Panhandle right now, when it gets to Austin, it will be a tragedy."

In the same article Austin Fire Department Battalion Chief Palmer Buck was even more blunt: "It would not be what house we are going to save . . . it would be what neighborhood are we going to save."

By the time the fires were out in Texas in mid-March 2006, the acreage burned to that point in fires in the United States stood at 930,000. That's ten times the average to that date, and nearly all of it was burned in Texas and Oklahoma. From December 1, 2005, to March 15, 2006, there had been 8,900 wildfires in Texas, with 85 percent of them starting within two miles of a town or city.

These firestorms in Texas pointed out an ominous vulnerability of people to the effects of fire. Although it was widely reported that ten thousand cattle and horses had died in the flames, the bigger, underreported story was about many of the cattle that lived.

The fires in the Texas Panhandle not only destroyed fields of grain and hay over a large geographical area; they also burned the hay that ranchers had stockpiled to feed their cattle. As the fires moved through, many ranchers whose cattle did survive were faced with the choice of bringing them quickly to market because the ranchers no longer had any way to feed them. This could have been a frightening scenario if it had happened in one of the many areas of the world that rely only on local food sources.

Residents of the Panhandle were fortunate that five days after the firestorms began the flames were doused by a weather system that brought bands of rain and sleet. The same system caused flooding in the Dallas area that killed at least one person. It seemed to Texans that they were facing one extreme or another.

Despite the rain, officials were not confident that the drought would be over anytime soon. In fact, some warned that if East Texas remained dry through the spring it would be very bad news for the Chicago area, because the moisture that city far to the north relies on from the Gulf for its summer rainfall would have a long way to go over parched ground to get there.

Some officials in March 2006 were saying that the fire season in the desert Southwest in 2005 essentially never ended and feared that it would simply continue on through the next year. Tucson had experienced its driest winter on record and it had not rained at all in Phoenix for four months. Large range fires burned through New Mexico in November and December, during a time when fire season should have been long past. In February a four-thousand-acre fire burned through the Tonto National Forest in Arizona, and then in March a twenty-six-thousand-acre fire forced the evacuation of one community in New Mexico.

The earliest fire restrictions ever were posted this year in Arizona, and before winter was even over fire lookout towers had been staffed and fire officials were holding community meetings. Some residents were demanding that all national forest land in Arizona be entirely closed to the public.[2]

Meanwhile, back in the San Bernardino Mountains, by late January hardly any precipitation had fallen since the fall and the brush and trees were bone-dry. On January 23, 2006, humidity was at 8 percent and sustained winds of 40 miles per hour and gusts of 65 miles per hour hammered the ridges. Firefighters were stationed all along the Rim of the World Highway above San Bernardino, dreading and watching for signs of fire.

They didn't have long to wait. By afternoon, some residents of Running Springs were under a voluntary evacuation order after a wildfire was sparked in nearly the same area the Bridge Fire started in 2003.

Texas, Oklahoma, and the desert Southwest were not the only areas of the country in danger during the winter of 2006. On March 16, high wind and low humidity prompted the National Weather Service to issue a red flag warning for all of Virginia, the Washington, D.C., area, and most counties in Maryland. In Prince George's County, Maryland, alone, five wildfires had been sparked in the first half of March.

The Virginia Department of Forestry had a single-engine air tanker on standby in mid-March of 2006, the first time that has happened in more than a decade. On March 15, it was used to help extinguish a twenty-acre wildfire. From January 1 to March 15, firefighters had responded to 577 forest fires in Virginia and 265 fires in Maryland. Virginia forestry officials said that the number of fires in the state had increased 385 percent over the same period in the previous year.

This is not the sparsely populated western plains. The Virginia and Maryland region is one of the most heavily populated areas in the country, with extensive areas of urban/wildland interface. Its urban sprawl rivals that of Southern California.

In the third week of March, red flag warnings were also up for highly populated areas in Florida and Alabama. Scattered fires were burning in Florida, Alabama, and Georgia. This region also contains many areas of wildland/urban interface. The increasing number of these areas in which wildlands butt up against cities and towns amplifies the potential for disaster.

While officials in fire-prone areas such as Southern California continually cite the danger of fires moving from wildlands into urban areas, the opposite can also happen. In June 2005, a fire started in a house in western Morongo Valley, a rural area about twenty miles northwest of Palm Springs. The fire quickly spread from the structure to the surrounding desert grass and brush. By the time it was over, the blaze had destroyed six homes, threatened hundreds more and burned six thousand acres.

WHEN RAIN IS NOT A GOOD THING

As is often seen in areas such as the Southwest, ample rain can be a mixed blessing, causing disastrous flooding from fire-denuded slopes and even greatly increasing fire danger. The case of Texas in 2006 is ironic. In addition to the wind and dry weather, proper land management by ranchers and two good growing seasons in the previous year actually contributed significantly to the magnitude of the disaster because they increased the fuel available for the fires.

In Southern California in late October 2003, rain and early-season snow caught the region by surprise but helped firefighters get a handle on the fire disaster there. Firefighters who had been soaked in sweat in their turnout gear in triple-digit temperatures fighting the Old Fire in the San Bernardino Mountains a few days before were now shivering in their boots.

At the height of the fire, from ten to twelve strike teams, which is approximately fifty to sixty engines, were fighting the blazes in the western San Bernardino Mountains. Five days into the firefight, the forecast was calling for snow all the way down to the forty-five-hundred-foot level, and the strike teams that were assigned to assist the local department did not have tire chains.

Crest Forest Fire Protection District Fire Chief William Bagnell in Crestline scrambled, with the help of a local auto parts dealer, to get snow chains up the mountain so that his visiting strike teams who did not have chains could stay and ensure the fire was completely out.

In the immediate term, the rain and snow had been a good thing. It was light, soaking precipitation and did not fall heavily enough to ravage the recently denuded landscape. It was exactly what was needed.

Two months later, the rain was not welcome.

SEVENTEEN

Disaster Begets More Disaster

Even though the Old Fire was out, it was not through causing great destruction and killing people.

Just before Christmas 2003, a series of wet storms began piling up on a subtropical jet stream in the western Pacific with a direct bead on Southern California, where many of the mountain slopes had been completely denuded by fire. Known by meteorologists as the "Pineapple Express," this phenomenon can dump huge amounts of rain on California.

In some years the Pineapple Express is welcome, because it can erase a drought in a week. But this one's timing could not have been worse.

The fire-denuded slopes of the San Bernardino Mountains rise in just a few short miles from fifteen hundred feet to as high as fifty-seven hundred feet above Waterman Canyon, where the Old Fire had burned through exactly two months before. The ground was still bare; there had been no time for new vegetation to grow and put down roots and ground cover to limit erosion.

Steep mountain slopes such as these are highly vulnerable to landslides and mud slides when heavy rains follow a fire such as the Old Fire, which completely destroyed the once thick vegetation in many areas along the southern slopes of the mountains.

In Waterman Canyon, where the Old Fire had started and several homes burned, the Greek Orthodox parishes from the Los Angeles area ran St. Sophia Camp, a streamside summer retreat. Motorists ascending Highway 18 can see it off to the right in a deep gully.

The camp had been spared the wrath of the fire, but it would become the site of a disaster and awful human tragedy just two months later.

On Christmas Day 2003 the heavens opened with a fury and at least sixteen straight hours of heavy downpours hit the steep, blackened southern slopes of the San Bernardino Mountains, which rise more than three thousand feet straight up from St. Sophia Camp, and a voluntary evacuation order had been given for Waterman Canyon residents.

The slopes had already been soaked before this latest deluge, and law enforcement and Caltrans officials knew Highway 18 through the canyon would probably be washed out.

San Bernardino County sheriff's deputies went door-to-door in the Waterman Canyon area on Christmas Eve, warning residents to voluntarily evacuate. Unfortunately, officials didn't know there were people at St. Sophia Camp.

The caretaker of the camp, which was closed at this time of year, had invited a group of extended family and friends to stay there. The operators of the camp would later say that they had no knowledge that the group was there and that the caretaker had not been authorized to let people stay there. Both the caretaker and his wife would be dead before Christmas was over.

In the early afternoon of Christmas Day, San Bernardino County Sheriff's Deputy Tracy Klinkhart saw a sight he will never forget. Klinkhart was driving back up Highway 18 toward Crestline and was approaching the first turnout just past the interchange with Waterman Canyon Road. "At about two o'clock I came across two adults and two kids without their shoes and caked with mud," Klinkhart said. "They crawled up here and kind of ran me through what happened. They said they had crawled out of the mud slide and were telling me there were possibly fifteen people still left down in it. They said they also saw some of them swept away."

At the same time Klinkhart found the survivors, a desperate call was made from the camp for help, and emergency personnel began trying to get into the area. The crews were thwarted in their efforts to reach the area by rock slides, mud slides, rushing water, and damaged bridges. A gas main in the road had ruptured, and natural gas was escaping at a high rate.

Fourteen people were rescued in the initial efforts on Christmas afternoon and night. Only body recoveries took place after that.

An hour before, farther upstream at Pohl Ranch, a 114-acre apple orchard in the canyon above the camp, caretaker Doug Johnson was struggling to divert the flows, which were creating waterfalls along his road. "Everything that hit the camp came through us first," Johnson said. The road was lost, but the house in which he lived with Theresa Kugle, also a caretaker, was spared.

Johnson had been using a backhoe, preparing for the worst. "We diverted the flow around it," Johnson said. "We've been doing a lot of work in the culverts and everywhere." The work almost cost Johnson his life. At the time of the flash flood, he was working in the backhoe, struggling to shore up the defenses around the property.

"It was moving boulders the size of cars," Johnson said. "I was working in the backhoe and all of a sudden I had rocks coming through the cab." Kugle and her daughter watched as Johnson jumped from the backhoe. When they got back to the house they realized their truck was stuck in the road. They called for a tow truck, but the tow truck could not make it through to the site. Finally, a fire truck came and pulled the truck out. Johnson parked the truck on Highway 18 above the house.

Johnson said the major flash flood event took place in approximately five minutes. The day after the flood, he and Kugle said they still had no idea where the backhoe was or how the orchard, which survived the fire, had fared.

Johnson said the flood was even too much for his preparations. "I had an eight-foot-deep culvert with six feet of dirt piled on top of it," Johnson said. "I never even imagined anything like that."

DOES FIRE BEGET MORE WARMING?

One disturbing aspect of wildfires caused by climate warming is what some scientists call the "feedback loop" of CO_2 release and uptake in both the boreal forests that ring the Northern Hemisphere and the tropical forests of the Southern Hemisphere.

Johann G. Goldammer, who leads the Fire Ecology Research Group at the Max Planck Institute for Chemistry, in Freiburg, Germany, has warned that fire caused by global warming could very well actually accelerate global warming.

"Drought and fire may also release carbon from the peat bogs and swamps of the boreal zone," Goldammer wrote. "Between 66 and 98 billion tons of carbon are estimated to be stored in living and dead plant biomass in the global boreal forest area. Thus more fires, arising from more frequent and intense drought, may produce an additional pulse of CO_2 to the atmosphere, acting as a feedback loop in global warming."[1]

Because the boreal and tropical forests absorb vast amounts of CO_2 in the atmosphere, many scientists say their destruction from fire, drought, and clearing will mean that far less CO_2 will be absorbed by the diminished forests, resulting in a potential feedback loop that could accelerate warming beyond current predictions.

Source: 1. Johann G. Goldammer, *Our Planet*, vol. 9 (New York: United Nations Environment Programme, 1998).

Despite a valiant rescue effort and weeks of searching, nine children, ages six months to sixteen years, along with five adults, were carried away and killed by the flood. Some bodies were found miles downstream from the camp. Two other people were killed in flash floods in Devore, ten miles to the west, at a KOA campground on the same day.

The severity of the Christmas Day event, which was first called a mud slide, was evident in Waterman Canyon, especially in the area of St. Sophia Camp, where the greatest devastation occurred. Throughout the camp and downstream from it, the many new watercourses that had been formed by the torrent were choked with various lengths of tree trunks burned in the fire, and boulders the size of stoves and refrigerators were strewn across the scene.

Shortly after the incident, Sheriff's Deputy Chip Patterson was asked whether the Waterman Canyon flood victims, along with the two people killed in Devore, could possibly be linked with the Old Fire and charges against them added to charges of murder against any arsonists who might be brought to trial. He said that he was not aware of any discussions in that regard at the time. He said it would depend on whether the sheriff's department felt its case linking the two events was strong enough to present to the district attorney's office. However, Patterson said he was doubtful such a case could be made.

Patterson said the disaster should actually be referred to not as a mud slide but as a massive flash flood—an event that occurred with little warning in a matter of minutes and in which mud slides were only one component. As Johnson related, the actual flash flood event took place in a matter of approximately five minutes, while some said that it seemed to happen in the space of a minute or less.

The Waterman Canyon event was a flash flood on a massive scale, with hillsides collapsing and combining with walls of water, pushing along trees, car-sized boulders, bridges, buildings, and anything else in their path. The torrent scoured the lower reaches of the canyon from its highest elevation on the mountain to the valley below. Motorists driving up State Highway 18 today can still see scars from the mountain giving way on that fateful Christmas.

Again, precipitation is not always a good thing, and not only because of the potential for flooding after a major fire. In fact, it often creates more fire danger. As spring began in 2006, Phoenix residents were looking to the sky, hoping for even a stray shower to end the dry spell. However, Bureau of Land Management officials did not share that hope. Record rains of the year before had created thick vegetation in Arizona, and officials said that a few small storms now would serve only to stimulate new growth, rather than filling the existing vegetation with much-needed moisture.

"What you need is not just one storm," said Rick Ochoa, national fire weather program manager for the U.S. Bureau of Land Management in Boise, Idaho. "You need a number of storms occurring over a month or two to really put a big dent in the fire season."[1]

This was the case in the mountains of Southern California after the Old Fire. Despite record rains and flooding just months later that closed mountain roads and even killed dozens of people, Southern Californians were looking forward to a lush green spring and summer.

They soon learned to be careful what they wished for. The green spring and summer came, even in the desert, where a display of greenery occurred like no other in recent memory. Lake Arrowhead in the San Bernardino Mountains, which normally receives thirty inches of precipitation in a good year, received more than seventy inches of precipitation during the 2003–2004 rainy season, which is measured from July 1 to June 30.

By August 2004, all of that dense, new growth was brown and officials were once again warning residents to clear their properties of combustible vegetation and materials and to be prepared to evacuate if given the order. Many in the area whose homes did not burn in the Old Fire began talking about remaining in the mountains the next time an evacuation was called, a move that could hamper future firefighting efforts and put thousands of lives at risk.

THE DAMAGE HAS BEEN DONE

As was seen in the San Bernardino Mountains, people were attached to the beauty of their forest homes and were not inclined to take the advice of ecologists and cut down some of the trees on their property to thin them out. In many cases, people were prevented from doing so by ordinances. The folly of that lack of forest management is seen today on the slopes northwest of Lake Arrowhead, which are filled with hundreds of pricey homes, some of them in the multimillion-dollar range.

In the year 2000, few of these homes were visible under

the thick forest canopy. By 2003 the canopy was gone, after bark beetle infestation caused nearly 100 percent pine tree mortality in the area. It was a blessing in disguise. If the dry forest had been there in October 2003, embers from the Old Fire that blew through there would have likely ignited the trees and destroyed the entire area.

Many believe that a forest is just that—many closely spaced trees that create a thick, green canopy. Most ecologists and forestry experts see such a forest as a human-caused tree population explosion and a disaster in the making.

In a paper in the *Journal of Forestry* titled "Southwestern Ponderosa Forest Structure and Resource Conditions: Changes Since Euro-American Settlement," W. Wallace Covington and Margaret M. Moore explained the public's difficulty in getting a handle on this concept: "Although most of the public is familiar with the consequences of deer population explosions resulting, in part, from overly zealous predator control, many fail to see the analogy to pine tree population explosions as a result of Euro-American settlement."[2]

When people go "back to nature" and enjoy many of the forests of the southwestern United States, many of them believe they are looking at the wilderness as it appeared hundreds of years ago. This is far from the reality. In most places, the ponderosa, or "yellow pine," forests of the Southwest bear little resemblance to the forests that existed in this area before Euro-American settlement.

One might believe that before technology allowed for active fire suppression vast areas and whole communities would have been wiped out by fire. However, fire itself was what kept such disasters from occurring in earlier times.

Native Americans all over North America frequently burned land without causing great harm to the environment or their communities. On the Great Plains they often used fire to improve grazing lands. They knew that fire served to reinvigorate the prairie and that bison loved to feed on re-growth in burned areas. So Native Americans often used con-trolled burns. Native peoples who lived in forested areas used fire to burn undergrowth that limited their movement and also used it to reinvigorate habitat area for wildlife, which tends to decline in the absence of fire. This was true not only in the West but in forests throughout the country. Lightning also played a big part in this process.

It has been estimated that Native Americans burned an average of 13 percent of the land in California every year in the time before Euro-American settlement. These fires not only improved ecosystems but also likely limited the number of gigantic wildfires. Active fire suppression has actually in-creased the size of fires, according to many experts. From 1970 to 1992, California experienced a doubling in acreage burned by wildfires, while the number of wildfires in the state increased only slightly during the same period.[3]

Large increases in burned acreage of separate fires have also occurred in recent years in Idaho, Montana, Oregon, Washington, and Wyoming.

By the mid-1800s in many areas of the country, settlers had long discouraged and suppressed fires. This resulted in increased fuel loads in forests and prairies, and the composi-tion of the land changed greatly, making it highly susceptible to wildfire. At the same time, a far greater number of poten-tial human ignition sources were present because of settlement. In the 1880s, these conditions caught up with man and resulted in disastrous fires for years that caused people to begin to rethink fire suppression policies.

Even with all of our firefighting resources, forest fires today in the Southwest are much more likely to burn faster, last longer, and destroy more acreage than back in the days when people did not extinguish them.

Before Euro-American settlement the forests of the Southwest contained large, more widely spaced trees. There were far fewer young trees and far less underbrush. There was more grass growing below the trees, and open meadows were abundant. The appearance of the forests before settlement has been described as "parklike."

In the *Journal of Forestry* study, the authors explained how these forests now appear and what got them to that condition. "Heavy logging, and fire exclusion, along with climatic oscillations and elevated atmospheric CO_2, have led to many more younger and smaller trees; fewer older and larger trees; accumulation of heavy forest fuel loads; reduced herbaceous production; and associated shifts in ecosystem structure, fire hazard and wildlife habitat."[4]

What does this mean for fire hazard? One need only look at these two different descriptions of the forest to imagine how differently fire would behave in them. In the forests before Euro-American settlement, fires would often burn the grass through these widely spaced trees and through the plentiful meadows between them. The larger trees were much more resistant to fire.

This means that the "crown fires" we see in today's forest from fires that are conducted from the ground to the canopy would have been far less common. Also, because trees were more widely spaced, this type of fire would have spread more slowly. Crown fires also often cause total destruction of stands of yellow pine when they move through an area.

Today's forests contain many more younger trees that are spaced closer together and what fire officials call "ladder fuels," starting at the ground with grass and pine needles, going to dense brush, then up to the branches and crowns of closely spaced trees. Fires can burn across the crowns of trees with alarming speed, especially when they are weakened or killed by drought or infestation.

Crown fire is the type that raged through Cedar Glen during the Old Fire in a matter of hours, destroying more than three hundred homes.

Crown fires are particularly devastating. The Fridley Fire of 2001 in Montana was one of these. The fire burned for days, but at its beginning it was terrifying, burning fifty thousand acres in just six hours. That is seventy-eight square miles—nearly the same size as the area that the Baltimore City Fire Department covers, and three and a half times the size of Manhattan.

It was fortunate that the Fridley Fire happened in an unpopulated area, given the destruction described in the following account by writer and search and rescue volunteer Douglas Gantenbein, who said at the time he was witnessing the annihilation of a forest:

> From my vantage point I could easily see a column of smoke climbing like a giant mushroom from the mountains rising over Paradise Valley, 30 miles north of Yellowstone National

Park. The column rose with such fantastic energy that even from six and seven miles away it could easily be seen boiling upward, pushing higher and higher in swelling folds of hot gas and ash. The source of this awful sight was a firestorm ripping through a stand of dry, 150-year-old lodgepole pine trees. Lashed on by high afternoon winds, the fire was moving steadily northward from its starting point, its flames—as tall as a 25-story building—devouring trees, deer, elk, squirrels, cows, and everything else that could not get out of its way in time. Streams filled with mountain trout were reduced to steaming, moist strips; granite boulders heated to the point where they were boiling."[5]

WHEN WILDFIRE RIPS THROUGH POPULATED AREAS

The Fridley Fire described earlier tore through an uninhabited area. But what happens when fire tears through an area where people have their homes? The San Bernardino National Forest is the most urbanized national forest area in the country, but many areas that abut national forests and BLM land are becoming the same way.

Cedar Glen in 2003 was a mostly bedroom community in the San Bernardino Mountains along a scenic creek and covered in pine trees. From the heights around the area very few of the 550 homes were even visible because of the trees. Some of the trees here had suffered bark beetle damage, but overall, this well-watered canyon had not suffered as badly as other areas. Early in the twentieth century the area had sprung up as mostly a weekend camping destination, so most of the lots were closely packed and many of them would not be buildable given the county codes of today. The water system in the area was also out-of-date.

On October 29, 2003, during the Old Fire, nearly all of the five hundred-plus homes in Cedar Glen had been evacuated days before. If public education and awareness had not resulted in such a complete, cooperative exodus, the death toll might have been staggering. The experience at Cedar Glen is a lesson for other scenic towns everywhere around the country that are covered in a lush, green canopy.

Arrowhead Manor Water Company General Manager Darel Davis, appointed by the court to oversee the beleaguered company when it entered receivership, faced immense challenges in Cedar Glen, one of the oldest established areas of the mountain communities. A large number of the homes in his service area—more than 300 out of 550—were destroyed in just a few hours on that fateful October day. Although the former owners and managers of the district had been under criticism both before and after the fire, Davis was widely commended for the work he did to keep water going in the area during the fire.

Because of the power outage during the fire, Arrowhead Manor Water Company's only source of water was Crestline–Lake Arrowhead Water Agency (CLAWA), which supplied mostly Crestline and a few other areas of the mountain. CLAWA hooked up an emergency tie-in to the Arrowhead Manor Water District during all of the fires to keep the Cedar Glen area supplied with water.

On the day the fire started, Davis started preparing for

the worst. "This was our first disaster, so my crew and I had no experience with something like this," Davis said. "So we filled all of the tanks with water, and then we got the news that they were going to cut power. We overpressurized the system to see that we had plenty of capacity for the onslaught of the fire. In doing that we started having some problems out at the end of the system."

On the day the fire hit Cedar Glen, Davis and his son Brett went out to the Hickory Circle area at the end of the system to fix the main breaks that were occurring there, which they did before they were forced to flee the area, according to Davis. They barely made it out alive.

"We were there when the fire breached the canyon and got onto our ridge," Davis said. "The fire was going about forty miles per hour. We were driving up the road as fast as we could go and it was passing us. The fire sounded like you were standing behind a 747 floored. I never heard anything like that in my life."

After the fire swept the area, rumors abounded that there was not enough water pressure in the water company's system to feed the hydrants for firefighters to fight the fires. However, according to Davis, at the time the fire began sweeping rapidly through the canyon there was adequate pressure. "At that point all of the tanks were full and all of the system was in full operating condition," Davis said.

That soon changed as the fire converged from three different directions, destroying hundreds of homes and much of the infrastructure of the water company. It quickly destroyed a main gravity tank and cut off water service to the entire lower area that was served by it. As the fire progressed through the houses, the "bleeding" of the system intensified as individual services in homes were destroyed.

Shortly after, the fire destroyed a major tank and pump station, the pump of which had already been disabled by the power outage. "It's safe to say that once the fire reached the top of the ridge, most of those homes were gone and there wasn't much left of our system," Davis said.

During the evacuation, he was maintaining the system but also checking for looters and checking that people had gotten out.

Davis said the people of CLAWA and the other water agencies are to be commended. "These guys are unsung heroes," he said. "They got their families off the mountain and then they came back. Nobody was worried about their houses; nobody was dragging hoses around and wetting down their roofs. They were running around filling tanks and making sure everyone had the equipment and water they needed to do this fight. It was an amazing show of courage. I'm not taking anything away from the firefighters because I know what they had to do, but they would have had a hard row to hoe if the hydrants had been dry, and they weren't dry because of these guys."

Cedar Glen is a lesson for many forested communities all over the world that are susceptible to wildfire. Citizens must not become complacent about problems with their water supplies, and they must recognize that a more natural forest, with meadows, widely spaced trees, and relatively clear forest floors, will be much less likely to spread deadly and destructive fires.

They must also support efforts to prevent fires from happening in the future, and this includes thinning trees on their own properties and supporting the efforts of politicians and local fire agencies to prevent fire disaster. This is often easier said than done, especially when efforts to avert disaster in advance end up causing it.

Even when the conditions are not optimal, fire can easily get out of hand. On February 8, 2006, a prescribed burn got away from USFS firefighters in the Cleveland National Forest of Southern California. The fire burned about eleven thousand acres and forced the evacuation of two thousand residents of urban Orange County and caused the closure of five schools and three freeways. Fire crews from all over California responded to the blaze, known as the Sierra Fire. Eight people were injured.

Prescribed burns are normally conducted when humidity is relatively high and winds are calm. During the previous two years, several prescribed burns had jumped containment in Southern California, two of them in the San Bernardino Mountains.

On March 25, 2004, snow was on the ground in the ski resort town of Big Bear in the eastern part of those mountains when USFS personnel set a prescribed burn above the town. By the time the planned 500-acre fire had burned only 150 acres, winds unexpectedly picked up and the fire was out of control. Two popular Southern California ski resorts were evacuated, and people in two nearby towns were advised by sheriff's deputies to be prepared to evacuate. One hundred and fifty loud, angry residents faced Forest Service officials the next day, demanding answers and railing about mismanagement of the forests over the past twenty years.

EIGHTEEN

Global Warming?

The lungs of our planet are dying. Our tropical rain forests have the capacity to help heal the human damage of global warming, but at the same time they can be one of the great contributors to accelerating its progress.

In 1997 and 1998, during the twentieth century's strongest recorded El Niño event, droughts in the Amazon and Africa caused widespread moisture depletion in soils and plants. In those drought areas, more than fifteen thousand square miles of standing Amazon forests were decimated by wildfires. On the African continent during that time, nearly eight thousand square miles of tropical rain forests burned. These fires in South America and Africa not only destroyed large areas of carbon-absorbing forest, but themselves also released huge amounts of carbon into the atmosphere, by some estimates as much as 5 percent of the annual amount of emissions produced by human activity.[1]

In 2005, it began happening again, and this time it was worse. In some areas of the Amazon in 2005, the drought was the worst seen since records started being kept more than one hundred years ago. Wildfires raged throughout the drought-stricken areas and villages were cut off from the world as their watery "highways" and "roads" dried up.[2]

In Acre State, one of the hard-hit areas, wildfires tripled to about fifteen hundred compared from the previous year. The smoke from the fires themselves was said to have exacerbated the situation by preventing the formation of rain clouds, which might have served to extinguish some of the fires.

Scientists say that as the oceans continue to warm, droughts in the Amazon and Africa will become more common. Scientists on ocean warming have blamed the 1997–1998 droughts squarely—and other scientists are blaming the 2005 drought—on warmer waters in the Atlantic. Researchers led by Daniel C. Nepstad of the Woods Hole Research Center in Massachusetts and the Amazon Institute of Ecological Research concluded: "The increase in forest flammability associated with severe drought poses one of the greatest threats to the ecological integrity of Amazon forests."

The year 2005 was a disaster. Literally. The United Nations Environment Programme (UNEP, www.unep.org) reports that 2005 was the costliest year ever for atmospheric-related disasters. The Munich Re Foundation estimated that weather-related disasters in 2005 caused $200 billion in damage, with insured losses at $74 billion.

This significantly eclipses the year before, which was the second most costly year, with $145 billion in damage, and $45 billion in insured losses.

UNEP pointed the finger squarely at global warming, blaming it for an increase in both number and intensity of damaging hurricanes and cyclones.

Anyone who doesn't believe in global warming should consider what recently happened to residents of one community in Tegua, one of the northern provinces of the Pacific island chain of Vanuatu. The United Nations is calling them the world's first "climate change refugees." This community became the first to be relocated because of the effects of global warming. The village was moved higher up in the interior of the province in 2005 after homes in their ancestral village were repeatedly swamped by storm surges and aggressive waves.

"The peoples of the Arctic and the small islands of this world face many of the same threats as a result of climbing global temperatures, the most acute of which is the devastation of their entire ways of life," said United Nations Environment Programme Executive Director Klaus Toepfer.[3]

"The melting and receding of sea ice and the rising of sea levels, storm surges and the like are the first manifestations of big changes under way which eventually will touch everyone on the planet. The plight of these vulnerable peoples should be a clear signal to governments meeting here in Montreal that we must hurry up if we are to avert a climate-led catastrophe for current and future generations," he added.

LESS FIRE, MORE DAMAGE

Changes in fire behavior in the United States are typical of what is happening all over the world. The most alarming change is the great increase in major fires.

In 2005, more than 8.68 million acres, or 13,573 square miles, burned in wildland fires throughout the United States, breaking the previous acreage record set in 2000. By mid-September 2006, officials with the National Interagency Fire Center (NIFC) in Boise had reported that the 2005 mark had already been broken, as fires continued to rage. By the middle of October 2006, 9.3 million acres, or 14,531 square miles, had burned in 84,884 fires, with more than two months left in the calendar year. But that astonishing figure pales when in the year 2007 it was surpassed again by the worst U.S. fire year in recorded history: nearly 10 million acres burned. And while most would assume the huge Malibu and San Diego fires were to blame for this frightening statistic, it was Idaho that suffered the most acreage burned with 1,980,552, compared to California's 1,087,110.

This means that the three most devastating years of wildland fires in terms of acreage since reliable records

These changes include an increasing danger of wildfire on a grand scale. In North America, increased lightning strikes and more frequent windstorms are expected. As temperatures warm, locations like Southern California that have historically seen wildfire will begin to see longer and hotter fire started being kept in 1960 have happened in this decade. Also, it is clear now that while the number of fires reported each year in the United States has been decreasing considerably, the amount of acreage is exploding.

I have gathered thirty-six years of NIFC data and averaged the numbers of reported wildland fires and their acreage. During the 1970s and 1980s, the average number of fires in the United States was 159,220 per year, with average acreage burned in each year at 3,715,202. Then things started heating up. From 1990 to 2005, the average number of fires per year dropped to 99,333, but the average acreage burned per year jumped to 4,823,413.

That translates to a 60 percent decrease in the number of fires, with a 30 percent increase in acreage. This means that the acreage per fire doubled. And keep in mind that in 2005 and 2006 the acreage per year nearly doubled again from the numbers in the 1970s and 1980s while the number of fires declined even more. These changes are far more drastic than can be accounted for by changes in fire management practices.

Source: National Interagency Fire Center, Boise, Idaho, http://www.nifc.gov/.

seasons. Forests in many places around the world will become drier and hotter. Areas in the more northern latitudes that have largely been spared the effects of wildfire will begin to see more frequent fires.

The process is already beginning. In 1997 and 1998, drought

caused by El Niño conditions was responsible for fires that burned large areas of forest all over the world. Another global outbreak occurred in 1999 and 2000. In the past ten years, Brazil, Ethiopia, Indonesia, Australia, Central America, the western United States, and the eastern Mediterranean have been hit particularly hard and with greater frequency. All of the European countries along the rim of the Mediterranean are susceptible to mammoth fires with the capacity for great human tragedy.

Wildfires can also create significant air pollution that can cause adverse health effects far from the source. The United Nations Environment Programme reported that smoke from 1997 fires in Guatemala, Honduras, and Mexico drifted over a large portion of the southeastern United States, prompting Texas officials to issue a health warning in that state.

This smoke problem was seen in the fall of 2003 in Southern California. During the 2003 fires I traveled all over the region and there were few places that were clear of choking smoke for long. The stench was so prevalent that it was not only outside but also inside cars and buildings. The smell got into furniture, carpets, and people's clothing even miles away from any flames.

NASA has estimated that annually humans burn up to 3 million square miles of forest and grassland around the world and they are responsible for 90 percent of all biomass burned, with naturally caused fires accounting for only 10 percent. A check of NASA satellite imagery in the third week of March 2006 showed widespread—and sometime massive-scale agricultural burning—taking place in West Africa, South Asia, and Southeast Asia. NASA says such widespread burning, while not immediately hazardous, "can have a strong impact on air quality and human health, natural resources, and climate."

Significantly, this burning, which has increased greatly over the past hundred years, releases large amounts of greenhouse gases, while at the same time diminishing the earth's natural capacity to soak up CO_2. This, in turn, causes the world to heat up.

Scientists have estimated that by 2100 the Earth will heat up another four degrees Fahrenheit overall. This may not seem like a big number to many, but consider the following: The difference between the average temperature of the Earth now and the average temperature of the last ice age, when vast sheets of ice covered most of North America, is only five to nine degrees Fahrenheit. Imagine the changes that are possible the other way with a 4-degree rise.

One final note: Just before I sent the final manuscript to the publisher, I attempted to once again reach the self-admitted arsonist I interviewed. The arsonist seemed so cocky the day I spoke with him and didn't seem to care about the pain his need—and according to every expert I spoke with, it is a need—causes so many people.

Like the dog chasing the car down the street, I don't know what I would have done had I known where to find him. But one imagines.

Each of us who survived the fires has a different story to tell. There could be hundreds of books written about the wildland fires of 2003, and even then still hundreds more stories would be left untold. The stories in this book represent my own and those of the very few I was able to locate during the research phase of *Hell on Earth*.

I'm tired of fire; tired of wondering what the siren screaming by on a hot afternoon means. We had the Big One twice in four years. Does the siren belong to the Next One?

As the writing of this book comes to a close, the dead trees continue to come down in Lake Arrowhead. But there are still countless thousands of dead trees—standing matchsticks, as Peter Brierty would say—ready to burn the next hot summer day some firebug wants to get his jollies. And arson notwithstanding, any kind of accident could spark a major fire, like the electrical accident that sparked the 2007 mountain blaze, consuming more than one hundred houses. Add the Santa Ana winds and we're right back where we were in October 2003 and 2007.

This book was a catharsis, defined as being "a purifying release of the emotions or of tension, especially through art." While this book probably doesn't qualify as art, writing it was a cleansing experience. So I'll leave it at that. I sincerely hope you never have to experience anything like the horrible firestorms recounted here.

Note to the arsonists reading this book: Get help. It's available. Please get help. Your predilection isn't worth the life of a child or that of the firefighter sent into harm's way as a direct consequence of your selfishness. Steven Rucker should be home with his wife and kids right now, as should the five brave firefighters who lost their lives in the Esperanza Fire in October 2006.

IT'S BEEN MORE THAN five years since the wall of fire came down Hook Creek Road and turned our home to ash.

We lost most of Gracie's baby pictures, a beloved pet, our computers and everything contained within them, and the boat I was building, the centerpiece of my plans to take Gracie fishing on the lake when she got a little older.

One sunny Wednesday morning, June 1, 2005, not even two years after the Old Fire, four houses burned in Cedar Glen—two totally destroyed—along Hook Creek Road, just yards from the location where our home burned. Investigators suspect arson. And so it goes.

THERE IS BLAME ENOUGH to go around for the tragedy that struck Southern California in fall of 2003. Some blame Mother Nature; some the arsonists and the careless; and others, the government. However, perhaps the biggest culprit was the enemy that San Bernardino County Fire Marshal Brierty feared the most early on as he tried to raise the alarm: public complacency and apathy.

Time simply ran out.

Once citizens and government officials sprang into action, great progress began to be made. But it was too late. The drought had gone on too long, the brush was too dry, and there were too many dead trees. The opportunists struck when the time was right, and the careless did their own damage. There simply wasn't enough time.

However, a few forward-thinking fire and law enforcement officials made a big difference. In the San Bernardino Mountains, for example, the greatest story is the one that never happened. Because of intense evacuation planning and getting the word out to everyone in the mountains, the evacuation of the entire area was orderly, quick, and nearly complete.

Less than a half hour after the Old Fire started in Waterman Canyon on the morning of October 25, 2003, a steady stream of cars was proceeding down Highway 330 and off the mountain, long before any evacuation was called. People just knew what to do and knew it was in their best interest to do it. By nightfall on the same day, the majority of the people in the western San Bernardino Mountains were gone.

For most residents, there was no questioning the logic of leaving, so they left. If thousands would have stayed, many of them likely would have died, and the entire area may have burned, because firefighting operations would have been seriously hindered. Unfortunately, many in these mountains are now saying that the next time it happens, they will stay.

The conditions that caused the fire disaster in Southern California are growing there again, despite the record rains of the winter of 2004–2005. In fact, those rains only served to speed up the growth of grass and brush below the mountain areas.

At the end of January 2006 precipitation totals for San Diego County were between 40 and 50 percent of normal for that point in the rain year, which is measured from July 1 to June 30. Although many beetle-killed trees have been removed in the San Bernardino Mountains, in February 2006 there were still large stands of dead trees in and around the town of Crestline.

These conditions are also building in other places around the United States. If the world continues to heat up, as scientists say it will, devastating wildfires like those that hit Brazil in 2005 could become common in places where they haven't been a problem in the past.

People should take a close look at everything that happened in California leading up to Hell on Earth. They need to band together in their communities for disaster preparation and find ways to head off disastrous fires. They also need to put pressure on government early enough to make a difference. Time ran out in California, and it is running out in other places as well.

The fires of fall of 2003 and 2007 will be remembered as the Big Ones. For now.

NINETEEN

Hell in 2007

Four short years. A little under fourteen hundred days is all it took for the horrific fires of 2003 to be eclipsed by firestorms so colossal that reality quickly surpassed imagination within hours of the start of the Malibu fire.

Firefighter Brent Hernandez had been on duty for thirty hours and was dog tired. The adrenaline he had been operating on until then was hurting his kidneys so bad that, with the added weight of his covers and his gear, he seriously feared he would pass out.

Then he saw the straw that for him broke the camel's back. There, a few hundred yards away, was a freestanding multimillion-dollar mansion completely surrounded by brush that likely hadn't been cleared since the house was built.

The scene filled him with rage, yet his professionalism won out. He pressed into service two other firefighters who had just finished laying a perimeter around a property across the street.

"Wynan, Smith, over here!"

They met at the mailbox—which still had mail inside it—and formed a plan to save the house. While Smith stood by with a hose at the ready, Wynan would touch off spot fires in the outer brush. The idea was to create at least some kind of fire break to protect the house, and the wind was favorable, blowing away from the house and toward the approaching fire.

If the wind kept up, that would delay the advancement of the fire line—at least for a while. In the meantime, Hernandez would sit in the nearby engine, and while he couldn't allow himself to fall asleep, he could at least rest his bones and rehydrate while monitoring the progress of the fire for his crew.

The fire had become an unstoppable force that had chased more than 500,000 people from their homes, and unless the shrieking Santa Ana winds subsided, and that wasn't expected for at least another day, they could do little more than try to wait it out and perhaps react—tamping out spot fires and chasing ribbons of airborne embers—to keep new fires from flaring.

"If it's this big and blowing with as much wind as it's got, it'll go all the way to the ocean before it stops," a volunteer who joined up with the Hernandez crew said, as sweat poured from around his goggles.

"We can save some stuff but we can't stop it."

Smaller fires began merging to form giant infernos, creating pillars of smoke that could be seen, they were told, from space, and shifting flames had already burned across nearly 600 square miles, killing two people, destroying more than 1,600 homes, and prompting the biggest evacuation in California history, from north of Los Angeles through San Diego to the border.

Triple digit heat was pushing the 6,000 firefighters to their limits. And even the much anticipated help from the air tanker corps was doing little to quell the flames.

"When they drop retardant, when they drop water, it's literally turning to mist," because the winds are so strong it dissipates, said Gary Franks of the California Department of Forestry and Fire Protection. It didn't help that the seventy-mile-per-hour Santa Ana winds made it nearly impossible to predict where the fire would go next. Just as crews prepared to make a stand, they would get outflanked by embers that hopped the lines and exploded into new fires. "You won't see a Santa Ana fire come down on you until it's too late," said Hernandez.

"We have had an unfortunate situation that we've had three things come together: very dry areas, very hot weather, and then a lot of wind," Governor Arnold Schwarzenegger said on one of hundreds of news reports that were keeping the media busy that day. "And so this makes the perfect storm for a fire."

Over a hundred miles from Hernandez and company, the fires were affecting some of California's celebrity residents by threatening more of Malibu, where many stars such as Mel

Gibson, Cher, Tom Hanks and Rita Wilson, Nick Nolte, Jennifer Aniston, Mel Brooks, Ryan O'Neal, and others have homes.

In Rancho Santa Fe, a suburb north of San Diego, houses burned just yards from where fire crews fought to contain flames engulfing other properties. And in the mountain community of Lake Arrowhead, cabins and vacation homes went up in flames with no fire crews in sight. "It was my home for forty years and now it's gone," said one resident.

The sweeping devastation was reminiscent of blazes that tore through Southern California four years before, and the ferocity of the Santa Ana winds forced crews to discard their traditional strategy and focus on keeping up with the fire and putting out spot blazes that threatened homes.

The Firefighters were especially concerned about the dense eucalyptus groves in Del Mar and Rancho Santa Fe, fearing that the highly flammable trees could turn neighborhoods prized for their secluded serenity into potential tinderboxes.

The usual tactic is to surround a fire on two sides and try to choke it off. But with fires whipped by gusts that have surpassed 100 miles per hour, that strategy does not work because embers can be swept miles ahead of the fire's front line.

In those cases, crews must keep ten to thirty feet back from the flames or risk their own lives. Any flame longer than eight feet is considered unstoppable, and even water and fire retardant will evaporate before they reach the ground.

In these situations the strategy generally is to fall back.

Firefighters will pick and choose their priorities in terms of what they can protect. Instead of trying to stop the fire, they must try to prevent it from burning resources. This is why the fire crews left the Green Valley Lake area near Lake Arrowhead. In truth, there was only one way in, and one way out, and the crews couldn't risk losing what little equipment they had to a wind-driven fire that could burst over a hill at any time.

In the suburbs north of San Diego, firefighters employed the same tactic as the Green Valley Lake crews used.

As fingers of flame pulsed across a ten-lane freeway and raced up a hill on the opposite side in just seconds, the firefighters retreated in order to save the gear. The fire engulfed white-washed homes at the top of the ridge.

Firefighters battling two fast-moving blazes in Lake Arrowhead were also taxed by steep terrain, winding roads, and a forest packed with dead or dying trees. More than two hundred homes burned in Lake Arrowhead and Running Springs before it was all over.

Weather conditions only grew worse, with temperatures across Southern California about ten degrees above average. Temperatures were in the nineties by mid-afternoon and wind gusts were up to 60 miles per hour and were expected to be higher in the mountains and canyons.

So it all sounds the same.

The usual cast appears onstage: the heroic firefighters, the bureaucratic double-speakers, the concerned governor (who ironically stood center stage the last time around in 2003), the ignorant home owners who have more brush surrounding their homes than sense in their heads, and the

Feds with the photo-ops. And every time it gets a little worse. The figures are staggering, yet not unexpected.

Since the last fires blasted through Southern California in 2003, very little has been done to mitigate these disasters. Home owners are apathetic to basic fire prevention tasks like the aforementioned brush clearance and fire escape plans, and government agencies are still mired in red tape, refusing to spend relatively small dollars compared to the billions that are required to fight the fires when they come. And as we saw in October 2007, they do come. Here is the butcher's bill from these last firestorms.

Acreage: 516, 356.

Homes destroyed: 2,013.

Deaths: nine directly due to fires; seven involving
    evacuees, including infirmity, age, accident.

Injuries: 71 firefighters, 27 civilians.

When will a collective mind-set change occur that will inspire efforts to, if not suppress future fire events, at least be better prepared for them when they do happen? And how much more can the Earth stand as the smoke and burn-off from these massive fire events continue to cloud the skies? Until we toss politics aside and become one with the fact that we have to do something now, not later, we are doomed to a repetitive past. How many fires will it take? Are we ready to prepare?

Mankind must first prepare for global warming by building resilient societies and fostering sustainable development, says a scientific collective in a report published in the journal *Nature*.[1]

The researchers say climate change is inevitable and policymakers should be planning adaptation strategies to minimize the negative impacts of future environmental stresses on society. They say new ways of thinking about, talking about, and acting on climate change are necessary if a changing society is to adapt to a changing climate. They believe the obsession with researching and reducing the human effects on climate has obscured the more important problems of how to build more resilient and sustainable societies, especially in poor regions and countries

Adaptation has been portrayed as a sort of selling out because it accepts that the future will be different from the present. Their point is the future will be different from the present no matter what, so not to adapt is to consign millions to death and disruption.

These scientists say mitigation alone is not a safe strategy for staving off the negative consequences of climate change.[2] As they see it, the key difference is that adaptation is the process by which societies make themselves better able to cope with an uncertain future, whereas mitigation is an effort to control just one aspect of that future by controlling the behavior of the climate.

Over the years policymakers and scientists have proposed a number of mitigation schemes, including sequestering carbon dioxide from the atmosphere, cutting greenhouse gas emissions, and reforesting large areas of land, but the scientific collective argues these steps are not enough and that sustainable development and adaptation must be integrated into any potential framework on climate change. To define adaptation as the cost of failed mitigation is to expose millions of poor people in compromised ecosystems to the very dangers that climate policy seeks to avoid.

By contrast, defining adaptation in terms of sustainable development would allow a focus both on reducing emissions and on the vulnerability of populations to climate variability and change, rather than tinkering at the margins of both emissions and impacts.

By introducing sustainable development into the framework, the collective forces us to consider the missed opportunities that for the past fifteen years or more have focused enormous intellectual, political, diplomatic, and fiscal resources on mitigation, while downplaying adaptation by presenting it in such narrow terms as to be almost meaningless.

The scientists believe that until adaptation is institutionalized at the level of intensity and investment at least equal to the United Nations Framework Convention on Climate Change and Kyoto, climate impacts will continue to mount unabated, regardless of even the most effective cuts in greenhouse gas emissions.

Not a pretty picture.

But change can be had. In fact, we have a history of being able to effect change pretty quickly when we want to.

Consider London, England, in the late nineteenth century. The air was nearly unbreathable because of airborne pollutants generated daily by burning tons of coal and from other sources. Yet we turned that around, and even third world nations are beginning to see environmental issues with an eye to the future. More rivers are being cleaned up, and with every ecological disaster such as a major oil spill or compromised water system, these countries are beginning to realize that such actions can no longer be ignored. So hope exists, and we may pull our heads out of the sand after all.

Epilogue

The California Fire and Rescue Mutual Aid System brought together more resources than had ever been used in its fifty-four-year history. This system effectively coordinated the response of approximately 5,480 personnel who staffed 1,160 local government fire engines and 102 OES fire engines.

California's neighboring states, Nevada, Arizona, and Oregon, also provided vital support to the siege by sending in 120 additional fire engines and the accompanying staff.

Combined local, state, and federal resources totaled 15,631 personnel from the fire services and 1,898 fire engines. It exceeded the previous wildfire record set by the wildland fires of 1993.

The multitude of people who experienced the wrath of October and November of 2003 each dealt with the fire and its aftermath in their own way. Some of the victims chose to rebuild; many did not.

My family decided against rebuilding. We used the eighteen months following the fires to regroup, renew, and reflect. Reflection took us to Oregon, where we now live in a house northwest of Eugene, near the TV studio where I produce the children's program *Nanna's Cottage*.

I left the *Mountain News* in June 2005, having enjoyed my position as the editor and news writer in Lake Arrowhead. To this day I continue to hope that some of the articles I wrote before the fires about the dead trees, and the need to cut them down, inspired at least some home owners to make their properties ready for the Big Ones. As we all know now, the Big Ones arrived with a vengeance.

PETER BRIERTY, fire marshal for San Bernardino County, is now an assistant chief for San Bernardino County Fire. In 2005, Chief Brierty was named the recipient of the National Forestry Heroism Award "for contributing countless hours of his personal time, in addition to his official duties, to preserve and protect the lives of hundreds of thousands of San Bernardino residents and tourists through large-scale fire prevention efforts."

LEE REEDER, our reporter and associate *Courier* editor from Crestline during the 2003 fires, is now the director of Inland Empire Waterkeeper, a grassroots, nonprofit water-quality organization. Lee launched a successful scenic photography business, www.1000wordsphotos.com, and writes and shoots photos for national publications. He also works as a freelance writer on my company's various television productions.

MICHAEL P. NEUFELD, the former *Mountain News* reporter, was promoted to editor of the *Mountain News* and *Crestline Courier-News* shortly after my family and I left Lake Arrowhead. He has since left that position and has resumed his successful freelance writing career.

THE FIREFIGHTERS portrayed in this book each remain with their respective agencies.

SERGIO MARTINEZ served his halfway house arrest and continues probation for his role in starting the Cedar Fire.

RAYMOND LEE OYLER was indicted for five counts of arson-related homicide and was charged with multiple counts of arson and arson-related activities surrounding the Esperanza Fire. He awaits trial.

JIM KOURI continues his career both in, and in support of, law enforcement and pro-law-enforcement causes. Kouri is a prolific writer who has become a strong voice in conservative issues in America and the world. He makes frequent guest appearances on radio talk shows.

THE RUCKER FAMILY continues to mourn the loss of Engineer Steven Rucker, as do his firefighting brothers and sisters in Novato.

GOVERNOR GRAY DAVIS was removed from office in a recall election that saw action movie star Arnold Schwarzenegger become governor of California. A landslide upon the completion of his first term reelected Schwarzenegger.

HARRY BRADLEY continues to publish the *Mountain News*.

THE HONORABLE ANN VENEMAN, SECRETARY OF AGRICULTURE resigned her Bush administration cabinet position following the president's reelection in November 2004. She now heads Unicef.

DAVID CAINE of the Arrowhead Communities Task Force accepted a position as field representative to State Senator Jim Brulte shortly after the fires. Caine continues to remain active in the fire safe council and is actively involved in the Lake Arrowhead community.

THE JIM JOHNSON FAMILY, who left their home and drove to the Disneyland Resort to wait out the fire, returned to find their home intact. They are thankful.

THE LICHTMANS of Simi Valley were also fortunate. Their neighborhood was left untouched by the flames of the Simi Valley Fire. David Lichtman continues his career in TV advertising and launched his own agency in 2007.

THOSE ARE JUST A few of the people who lived the story of the great California wildland fires of 2003. To everyone who shared personal memories with me, your stories are the true embodiment of what Hell on Earth was really like during the autumn of 2003 and again in 2007. And of course, sincerest condolences to those who experienced true loss. Houses and structures can be rebuilt. To those who lost loved ones, there is an unfair and nonnegotiable finality. For them, there remains the day before the fires and all the long days after.

ACKNOWLEDGMENTS

A book of this scope could never have happened without the cooperation of hundreds of people. I'd like to take a moment and acknowledge their contributions to this work and to sincerely thank them for all they did.

First, much love and thanks go to my wife, Karen, and our children, Richie, Jacob, Rachael, and Grace. They lived the horrible October of 2003 with me and helped me so much during the research and writing of the book.

In addition, many thanks and special appreciation to Lee Reeder, who began as a researcher and fact-checker on this book and who brought so much to it that I think of him as co-author. As always, Lee goes above and beyond. Thanks, Lee. Here's to the Dalmatian Coast, and all that comes from there. Cheers, buddy!

Thanks also to my good friend Steve Hodel, who introduced me to my agent, Bill Birnes, and freely shared his own writing and publishing experiences with me.

To Bill Birnes, I give a special thank-you for believing in the book and for believing in me as a writer. But I especially want to thank you for your dogged determination to place the project with the right publisher.

Big thanks go out to the proofing (and comedy) duo of Kate Osborne and Joan Barkdull: two very gracious and extremely sweet ladies. Girls, you are dolls!

To Tom Doherty and the team at Tor-Forge Books, thanks for bringing *Hell on Earth* to the world. Thanks, too, to editors Bob Gleason and Eric Raab. You guys definitely know what you're doing. And thanks to editorial assistant Melissa Frain for keeping the details ball rolling.

Special appreciation goes to the amazing Barbara Wild, to whom the moniker of copy editor is the understatement of the century. Barbara, your attention to detail, your excellent feedback, and your straightforward notes actually make me look like an author.

And finally, heartfelt thanks are given to the talented staff of the *Mountain News* and the *Crestline Courier-News*, especially news writers Michael P. Neufeld, Joan Moseley, and Harry Bradley, publisher.

APPENDIX

An Anatomy of What Occurred

Los Angeles County

Padua Fire: No deaths, 59 homes destroyed, 10,466 acres (16 square miles). Started October 22, 2003. Separated from Grand Prix Fire in San Bernardino County.

Verdale Fire: No deaths, 8,680 acres (14 square miles). Started October 24, 2003. Blamed on arson.

Riverside County

Mountain Fire: No deaths, 21 homes destroyed, 9,742 acres (15 square miles. Started October 26, 2003. Cause still under investigation, but investigators believe it may have been started by suspect Raymond Oyler who is in custody facing arson-related murder charges from the five deaths caused by the Esperanza Fire in October 2006.

San Bernardino County

Grand Prix Fire: No deaths, 52 homes destroyed, 69,894 acres (110 square miles) burned. Started October 21, 2003. Blamed on arson.

Old Waterman Canyon Fire (a k a: The Old Fire): Started October 25, 2003. Blamed on arson.

Final Statistics:

Acres Burned: 91,281 acres (143 square miles).

Communities Evacuated:
Apple Valley
Arrowbear
Arrowhead Springs
Baldy Mesa
Big Bear
Blue Jay
Cedar Glen
Cedarpines Park
Crest Forest
Crestline
Del Rosa
Devore
Green Valley Lake

Highland
Holcomb Valley
Hook Creek
Lake Arrowhead
Los Flores Ranch
Lucerne Valley
Oak Hills
Oak Springs Ranch
Rimforest
Running Springs
San Bernardino
Silverwood Lake
Skyforest
South Hesperia
Summit Valley
Twin Peaks
Valley of Enchantment

Deaths and Injuries: Six deaths, twelve injuries.

Firefighters Assigned at Peak: 4,211.

Structural Damage: 840 homes destroyed, 35 homes damaged, 10 commercial properties destroyed.

San Diego County

Cedar Fire: Start date and time: October 25, 2003 at 5:37 P.M. Human caused (accidental).

Final Statistics:

Acres Burned: 280,278 (440 square miles).

Administrative Unit: Cleveland National Forest/CDF San Diego Unit/Cedar Fire.

Conditions: One firefighter fatality and three firefighter injuries occurred in the Julian area on October 29, 2003. This fire was under a unified command USFS and CDF.

Control: Estimate for control was November 16, 2003, at 6:00 P.M.

Cooperating Agencies: CDF, U.S. Forest Service, local government.

Costs to Date: $27 million.

Deaths and Injuries: Fifteen civilian fatalities, one firefighter fatality, 104 firefighter injuries.

Location: Southern San Diego County.

Structural Damage: 2,232 residences, 22 commercial properties, and 566 outbuildings destroyed. 53 residences, and 10 outbuildings damaged. 148 vehicles destroyed.

Total Fire Personnel: 1,478.

Camp Pendleton: No deaths, 9,000 acres. Started October 21 on the Marine base, cause inconclusive, case remains open.

Dulzura Fire: No deaths, 46,291 acres (73 square miles) burned, 1 home, 5 outbuildings, and 11 structures damaged. Started October 26, cause believed to be stray, wind-borne embers from the Cedar Fire. Briefly burned across border into Mexico.

Otay Fire: The Otay Fire, which burned along the U.S.-Mexico border, charred more than 45,000 acres (70 square miles) in the South Bay area. That fire was fully contained by Wednesday morning, October 29.

Ventura County

Piru Fire: No deaths, 1 home, 6 outbuildings, and 1 commercial property destroyed, 63,719 acres (100 square miles). Started October 23, 2003. Cause never determined.

Simi Valley Fire: No deaths, 37 homes and 145 outbuildings destroyed, 107,590 acres (168 square miles; Ventura and Los Angeles Counties). Started October 25, 2003. Cause still under investigation.

NOTES

16. Hell Continued

1. A. Copelin and C. Osborn, "Catastrophe for Austin If Wildfires Hit," *Austin American-Statesman,* March 16, 2006.

2. A. Wagner, "Dry, Brittle and Stormless, the Southwest Is So Arid, Fire Season Is Expected to Arrive Early," *Washington Post*, March 16, 2006.

17. Disaster Begets More Disaster

1. A. Wagner, "Dry, Brittle and Stormless, the Southwest Is So Arid, Fire Season Is Expected to Arrive Early," *Washington Post*, March 16, 2006.

2. W. Wallace Covington and Margaret M. Moore, "Southwestern Ponderosa Forest Structure and Resource Conditions: Changes Since Euro-American Settlement," *Journal of Forestry* 92 (1994): 39–47.

3. R. E. Martin and D. B. Sapsis, "Fires as Agents of Biodiversity: Pyrodiversity Promotes Biodiversity," in R. R. Harris, D. E. Erman, and H. M. Kerner (eds.), Proceedings of the Symposium of Biodiversity of Northwestern California, Wildland Resources Center Report No. 29. University of California at Berkeley (1992): 150–177.

4. Covington and Moore, "Southwestern Ponderosa Forest Structure and Resource Conditions."

5. Douglas Gantenbein, *A Season of Fire: Four Months on the Firelines of America's Forests* (New York: Penguin, 2003).

18. *Global Warming?*

1. Daniel C. Nepstad, et al., "Amazon Drought and Its Implications for Forest Flammability and Tree Growth: A Basin-Wide Analysis," *Global Change Biology,* Vol. 10, Issue 5, (May 2004): 704–717.

2. L. Rohter, "A Record Amazon Drought and Fear of Wider Ills," *New York Times,* December 11, 2005.

3. United Nations Environment Programmes, *Global Environment Outlook*, vol. 3 (Sterling, Va.: Earthscan, 2002).

19. *Hell in 2007*

1. Roger Pielke, Jr., Gwyn Prins, Steve Rayner, and Daniel Sarewitz, "Climate Change 2007: Lifting the Taboo on Adaptation," *Nature* 445, 597–598 (February 8, 2007); published online February 7, 2007.

2. D. Sarewitz, head of the Scientific Collective, Arizona State University's Consortium for Science, Policy, and Outcomes, in a report released February 2007.

GLOSSARY

Anchor Point. An advantageous location, usually a barrier to fire spread, from which to start constructing a fire line.

Area Command. An organization established to: (1) oversee the management of multiple incidents that are each being handled by an incident management team (IMT) organization; or (2) oversee the management of a very large incident that has multiple IMTs assigned to it. Area Command has the responsibility to set overall strategy and priorities, allocate critical resources based on priorities, and ensure that incidents are properly managed, objectives are met, and strategies are followed.

Average Bad Day. Fire conditions experienced during typical day in the middle of the fire season. Used as a benchmark to gauge fire situations.

Backfire. A fire suppression tactic. Any intentionally set fire used to consume the fuel in the path of a free-burning wildfire.

BIA. Bureau of Indian Affairs.

BLM. Bureau of Land Management.

CALMAC. California Multi-Agency Command. The information coordination center established in Sacramento. Tasked to gather timely information from regions, cooperating agencies, the media, the director, interested government leaders, and the public.

CDF. California Department of Forestry and Fire Protection.

Chief Officers. Agency administrators, fire chiefs and other strategic level staff overseeing incident commanders.

Containment. A fire is considered contained when it is surrounded on all sides by some kind of boundary but is still burning and has the potential to jump a boundary line.

Controlled. A fire is controlled when there is no further threat of it jumping a containment line. While crews continue to do mop-up work within the fire lines, the actual firefight is over.

Convection Column. The rising column of gases, smoke, fly ash, particulates, and other debris produced by a fire.

Cooperating Agency. An agency supplying assistance including, but not limited to, direct tactical or support functions or resources to the incident control effort.

Crown Fire. A fire that advances from top to top of trees or shrubs, more or less independently of the surface fire.

Defensible Space. A fire-safe landscape for at least thirty feet around homes (and out to one hundred feet or more in some areas), created to reduce the chance of a wildfire spreading and burning through the structures. This is the basis for creating a defensible space—an area that will help protect a home and provide a safety zone for the firefighters who are battling the flames. In some states, it is required by law.

Direct Attack. A method of fire suppression in which suppression activity takes place on or near the fire perimeter.

Direct Protection Area (DPA). That area for which a particular fire protection organization has the primary responsibility for attacking an uncontrolled fire and for directing the suppression action.

Drawdown Level. The level at which the success of extinguishing a fire with initial attack forces is compromised.

ESF4. Emergency Support Function 4. A component of the National Response Plan developed for FEMA. A document that outlines different agencies' responsibilities in different types of emergencies.

ESRI. Environmental Systems Research Institute. A software company that produces software widely used to produce Geographic Information Systems maps of various emergencies for analysis and display.

Extreme Fire Behavior. "Extreme" implies a level of fire behavior characteristics that ordinarily precludes methods of direct control action. One or more of the following is usually involved: high rate of spread, prolific crowning and/or spotting, presence of fire whirls, strong convection column. Predictability is difficult because such fires often exercise some degree of influence on their environment and behave erratically, sometimes dangerously.

Federal National Team. A Type 1 National Incident Management Team coordinated by the National Wildfire Coordinating Group (NWCG). Team members may be from various agencies. The California Wildfire Coordinating Group (CWCG) sponsors five of the sixteen national teams.

Federal Regional Team. A Type 2 Incident Management Team maintained by the U.S. Forest Service in the Pacific Southwest Region (Region 5, California and the Pacific Islands). Team members may be from various agencies.

Federal Responsibility Area (FRA). The primary financial responsibility for preventing and suppressing fires is that of the federal government. These lands are generally protected by the Department of Agriculture, Forest Service, Department of Interior, Bureau of Land Management,

National Parks Service, U.S. Fish and Wildlife Service, and Bureau of Indian Affairs.

Fire Danger Rating. A management system that integrates the effects of selected fire danger factors into one or more qualitative or numerical indices of current protection needs.

Fire Line. A strip of area where the vegetation has been removed to deny the fire fuel, or a river, a freeway, or some other barrier that is expected to stop the fire. Hose lines from fire engines may also contribute to a fire being surrounded and contained.

Fire Perimeter. The entire outer edge or boundary of a fire.

Firescope. Firefighting Resources of California Organized for Potential Emergencies. A multiagency coordination system designed to improve the capabilities of California's wildland fire protection agencies. Its purpose is to provide more efficient resource allocation and utilization, particularly in multiple or large fire situations during critical burning conditions.

FMAG. Fire Management Assistance Grant. A federal assistance program managed by FEMA through the state Office of Emergency Services (OES). This program is designed to help state and/or local jurisdictions impacted by high-cost, high-damage wildland fires.

Fuels. Combustible materials.

GACC. Geographical Area Coordination Center; *see* South Ops.

GIS. Geographic Information System

Incident Commander. This Incident Command System position is responsible for overall management of the incident and reports to the agency administrator for the agency having incident jurisdiction.

Incident Command System (ICS). A standardized on-scene emergency management concept specifically designed to allow its user(s) to adopt an integrated organizational structure equal to the complexity and demands of single or multiple incidents, without being hindered by jurisdictional boundaries.

Incident Command Team (ICT). *See* Incident Management Team (IMT).

Incident Management Team (IMT). The incident commander and appropriate general and command staff personnel assigned to an incident. Also known as an Incident Command Team.

Indirect Attack. A method of fire suppression in which suppression activities take place some distance from the fire perimeter and offer the advantage of fire barriers.

Infrared (IR). A heat detection system used for fire protection, mapping, and hot spot identification.

Initial Attack (IA). An aggressive suppression action taken by first arriving resources consistent with firefighter and public safety and values to be protected.

Interface Zone. The area where the wildlands come together with the urban areas. Also referred to as the I-Zone.

Intermix Zone. The areas where homes are interspersed among the wildlands. Also referred to as the I-Zone.

Joint Information Center (JIC). An interagency information center responsible for researching, coordinating and disseminating information to the public and media. Formed through the Mountain Area Safety Task Force effort.

LRA. Local Responsibility Area.

MAC. Multi-Agency Coordination System, a combination of facilities, equipment, personnel, procedures, and communications integrated into a common system with responsibility for coordinating or assisting agency resources and support to agency emergency operations.

MAFFS. Modular Airborne Fire Fighting System (refers to the military aircraft C-130s, which are used as air tankers)

MAST. Mountain Area Safety Task Force.

MODIS. Moderate-resolution Imaging Spectroradiometer, a key instrument aboard the Terra and Aqua satellites. This instrument provided important intelligence for fire managers regarding fire perimeters and fire growth throughout the fire siege.

Mop-Up. Act of extinguishing or removing burning material near control lines, felling snags, and trenching logs to prevent rolling after an area has burned, to make a fire safe, or to reduce residual smoke.

Mutual Threat Zone (MTZ). A geographical area between two or more jurisdictions into which those agencies would respond on initial attack. Also called mutual response zone or initial action zone.

NIFC. National Interagency Fire Center, located in Boise, Idaho.

NPS. National Park Service.

OES. The California governor's Office of Emergency Services.

OSC. Operations Section Chief, the Incident Command System position responsible for supervising the Operations Section. Reports to the incident commander. The OSC directs the preparation of unit operational plans, requests and releases resources, makes expedient changes to the Incident Action Plan as necessary, and reports such to the incident commander.

Predictive Services. Those geographic areas and national-level fire weather or fire danger services and products produced by wildland fire agency meteorologists and intelligence staffs in support of resource allocation and prioritization.

Preparedness Levels. A national system of preparedness for incidents. The levels are 1 through 5. They are: *Preparedness Level 1*—few or no active fires under 100 acres. Minimal or no commitment of fire resources. Low to moderate fire danger. Agencies are above drawdown levels. *Preparedness Level 2*—numerous fires under one hundred acres. Local commitment of resources for initial attack. Moderate fire danger. Agencies above drawdown levels and requests for resources outside local area are minimal. *Preparedness Level 3*—high potential for fires one hundred acres and above to occur, with several zero- to ninety-nine-acre fires active. Fire danger moderate to very high. Mobilization of resources within the region and minimal requests outside of region. Agencies above or having difficulty maintaining drawdown levels. *Preparedness Level 4*—fires over one hundred acres are common. Fire danger is high to very high. Resource mobilization is coming from outside the region. Agencies at minimum drawdown levels. *Preparedness Level 5*—California Multi-Agency Command is fully activated. Multiple large fires are common in the north and/or the south. Fire danger is very high to extreme. Resources are being mobilized through the National Coordination Center. Activation of National Guard or military done or under consideration.

Red Flag Warning. Term used by fire weather forecasters to alert users to an ongoing or imminent critical fire weather pattern.

Rehabilitation. The activities necessary to repair damage or disturbance caused by wildfire or the wildfire suppression activity.

Santa Ana Wind. A type of Foehn wind that is a warm, dry, and strong general wind that flows down into the valleys when stable high-pressure air is forced across and then down the lee-side slopes of a mountain range. The descending air is warmed and dried due to adiabatic compression, producing critical fire weather conditions. Locally called by various names, such as Santa Ana winds.

Slop-Over. A fire edge that crosses a control line or natural barrier intended to confine the fire. Also called break-over.

South OPS. The multiagency geographic area coordinating center for Southern California. Located in Riverside, it is staffed by California Department of Forestry and Fire Protection, state OES, and federal fire agencies.

Spot Fires, or **Spotting.** A small fire ahead of the main fire caused from hot embers being carried to a receptive fuel bed. Spotting indicates extreme fire conditions.

State Responsibility Area (SRA). The California Board of Forestry and Fire Protection classifies areas in which the primary financial responsibility for preventing and suppressing fires is that of the state. The California Department of Forestry and Fire Protection has SRA responsibility for the protection of over 31 million acres of California's privately owned wildlands.

Strike Team. An engine strike team consists of five fire engines of the same type and a lead vehicle. The strike team leader is usually a captain or a battalion chief. Strike teams can also be made up of bulldozers and hand crews.

Unified Command. In the Incident Command System, unified command is a unified team effort that allows all agencies with jurisdictional responsibility for the incident, either geographical or functional, to manage an incident by establishing a common set of incident objectives and strategies.

WFSA. Wildland Fire Situation Analysis

Wildland/Urban Interface. The line, area, or zone where structures and other human development meet or intermingle with undeveloped wildland or vegetative fuels.

INDEX